THE *NUTS* *AND* *BOLTS* *OF* *NASCAR*

The Definitive Viewers' Guide to Big-Time Stock Car Auto Racing

Greg Engle

SPORTS
PUBLISHING

Sports Publishing books may be purchased in bulk at special discounts for sales promotion, corporate gifts, fund-raising, or educational purposes. Special editions can also be created to specifications. For details, contact the Special Sales Department, Sports Publishing, 307 West 36th Street, 11th Floor, New York, NY 10018 or sportspubbooks@skyhorsepublishing.com.

Sports Publishing® is a registered trademark of Skyhorse Publishing, Inc.®, a Delaware corporation.

Visit our website at www.sportspubbooks.com.

10 9 8 7 6 5 4 3 2 1

Library of Congress Cataloging-in-Publication Data is available on file.

Cover design by Tom Lau
Cover photo credit AP Images

ISBN: 978-1-68358-009-6
Ebook ISBN: 978-1-68358-010-2

Printed in the United States of America

CONTENTS

"There is no doubt about precisely when folks began racing each other in automobiles. It was the day they built the second automobile."
—Richard Petty

ACKNOWLEDGMENTS

TO THANK EVERYONE who has helped me during my time in and around NASCAR would require an entire volume. I do, however, want to single out several special people who were mentors and are now my friends. This includes Candice (Lee) and Reid Spencer, who got me my first NASCAR job. This also includes Bill Marx (the best editor I have ever known), whose numerous red pen marks on my early work taught lessons that I have carried with me and imparted to others throughout the years.

Also Kurt Busch and Todd Berrier—who are my "go-to" guys, the ones who explain things inside the sport that I don't understand. Also Richard Petty, who I once spent several hours with in his hauler while he explained—without his trademark cowboy hat and sunglasses—the inside workings of NASCAR.

Tony Fabrizio has been helpful through the years in allowing me to become a better reporter. Randy Fuller, the PR person for driver Carl Edwards, has been a wealth of information through the years and is someone I consider a good friend.

Every NASCAR driver I have interacted with over time has always been gracious and kind, especially when talking off the record. Specifically, Jimmie Johnson and Carl Edwards, who I worked for, for a short time. Both drivers are the type of guys I wish fans could see away from track. They are the type of people you would want to hang out with after work on a Friday night.

The one man who meant the most to me, however, was Jim Hunter. This NASCAR PR guru taught me more about NASCAR than anyone. We spent many, many hours together, and I learned more about the real workings of NASCAR from him than from anyone. He was the first person to call me after my father passed away in 2009, and he became like a second father to me. I miss him every day, and I will carry with me forever the lessons he taught me about NASCAR and life.

As for this book, all the people at NASCAR have been invaluable. Several NASCAR folks who stand out are Rosalie Nestore, Mike Forde, and Pete Stuart. Tireless Megan Englehart from Fox Sports spent many hours arranging interviews and getting photos. Jessica Rohlik from Joe Gibbs Racing and David Hovis from Team Penske were invaluable. Jessica secured an interview with driver Matt Kenseth and provided pictures, and David provided some wonderful shots of the Penske Racing facility. I am grateful to them all. Also thanks to Mike Zizzo from Texas Motor Speedway for the wonderful insights. In addition, thanks to another of my "go-to" guys, Larry McReynolds, champion crew chief-turned FOX Sports analyst, for taking the time to review this work.

My family has also been an invaluable source of support throughout the years. My children Amanda and Jon, both now grown, spent many days in their youth being dragged around a garage area and getting bored (and doing their homework) in the media center during NASCAR test sessions at Daytona, because Mom had to work. That "Mom" is my wife of 35 years, Carla, who has always been my rock. Then there's my own mom, Rose, who, thanks to her North Carolina location, has allowed me to save money on hotels through the years to cover races at Charlotte, Bristol, and Martinsville. I could not have done any of it without them.

Finally, I am grateful to all the people who call themselves fans of NASCAR. Without the first dollar they spend for a ticket, a sponsor product, or a piece of driver merchandise, none of it would be possible.

NASCAR is an industry and a lifestyle, because of its fans, and if not for the fans, thousands of us who make a living in the sport would be looking for jobs elsewhere.

Thank you, NASCAR fans—past, present, and future. It is to you this book is dedicated.

INTRODUCTION

AT SOME POINT you've seen it, either as part of a sports report inside your local news broadcast or perhaps you've been flipping TV channels and seen a bunch of brightly colored cars (or highly modified pickup trucks) racing around, usually in a circle. Maybe, at the time, you were curious enough to watch for a few moments, or perhaps until the very end, and wondered just what it was that you were watching.

Perhaps, all these days later, you are still curious to know more, or maybe you are a new fan lured in by a spouse or friend. No matter what has piqued your interest, what I hope you will learn in the following pages is just *what* is going on. A NASCAR race is much more than a bunch of race cars circling a track. To some it's a way of life, and to many it's the reason that they say, "Sundays are made for racing."

In the following pages, I try and break down all there is to know about the sport I first fell in love with as a young boy. In the early 1970s, my father, a professed IndyCar aficionado, took me to that most grand of speedways, Daytona International Speedway, while on vacation in Florida. The white sand beaches and tall stately palm trees were a far cry from my Indiana home. The race cars I saw at Daytona were different, too. Growing up, I journeyed to Indianapolis Motor Speedway almost every May, thanks to my dad's love of racing. There I witnessed greats like Mario Andretti and AJ Foyt. I was there in 1977 when Janet Guthrie qualified and raced in what was then the biggest auto race in America. The cars these men and women raced looked nothing like I saw anywhere outside a race track. They had no fenders, the cockpits were open, and the engines were in full view.

So it was then that I stood near the fence and looked in awe at the expanse of Daytona International Speedway for the first time. Indianapolis Motor Speedway was just as big, but it was hard to see just how big from the stands. Daytona was different. The 2.5-mile high-banked speedway stretched out before me like a vast desert with a ribbon of pavement surrounding it. That day several cars were testing, and I vividly remember that first car passing me. It exploded by with a roar that frankly scared the hell out of me. But I didn't just hear it, I actually felt it. The power, the speed, the sound of the engine as it rocketed off in the distance was something I wouldn't soon forget. These cars looked so different than the IndyCars I was used to. They had fenders, roofs, and hoods. In fact, they didn't look all that different from the cars I saw on the highway as we journeyed home to Indiana.

From that first brush with NASCAR to today, I have worked and lived around the sport. In previous decades, I worked as a member of the media at nearly all the speedways on which NASCAR races, and I now live not far from Daytona International Speedway here in Florida. Starting as first a fan, then a member of the media corps that endures the grueling travel schedule, I have enjoyed the sport as a fan and have also covered the sport as a reporter. I even spent a short time working as a PR person for a driver.

There are many people who watch some sort of professional sport. They may wear their team's jersey, sit in the stands, and cheer their favorite athlete during a game. Most of those games are familiar to all. As children, we probably played them, so we understand the basics: the ball goes in the hoop, crossing home plate equals a run, and crossing the goal line scores a touchdown. NASCAR is a bit different. There was no auto racing on the playground when we were released for recess, and few of us ever pulled on a five-point harness and helmet and drove as fast as possible for 500 miles.

Yet, NASCAR has become one of the largest spectator sports in the world—so much so that it's hard to escape it now. Some of the top consumer brands in the United States have a connection to NASCAR, so it's not unusual to see a display at the supermarket featuring a NAS-CAR theme, or a commercial on television featuring a driver or car racing around the track. On the weekends, while flipping through the channels, it's nearly impossible not to see a live NASCAR event somewhere in America. Whether it's qualifying, a practice, or the main event, the race itself is always seen live on a national network. If you live in town close to a racetrack visited by NASCAR, it's also hard not to notice the weeks they visit. Once, sometimes twice a year, the rolling circus known as NASCAR will invade, bringing thousands of people, temporary traffic problems, and general chaos to an otherwise quiet town.

In the following pages, I will give you enough information to allow you to understand just what is going on. Hopefully, you'll gain an appreciation of the sport of big-time stock car auto racing. Maybe next time you see it on TV, it will be less confusing, or maybe you'll get "it"—the same bug that has bitten millions of us—and someday you'll find yourself in the stands at a track as forty of the world's best stock car racers rocket past. Either way, my hope is that NASCAR will become less confusing to you. So welcome to NASCAR, or what those of us inside the sport sometimes refer to as the world of big-time stock car auto racing. My friends, pull them belts tight and hang on, because you are in for a wild ride.

CHAPTER 1

WHAT THE HECK IS NASCAR ANYWAY?

IT'S EASY TO tell a novice from a real fan. A novice will make perhaps the biggest mistake anyone can make when talking about NASCAR—they refer to a NASCAR race car as a "NASCAR." The stock cars raced in NASCAR aren't "NASCARs," though. They are actually highly modified Toyotas, Chevrolets, or Fords based on production models anyone can buy. While they cost millions of dollars to develop and build and could never be driven on the street, the intent is the same—to race cars that are akin to stock vehicles seen in dealer showrooms.

So just what is NASCAR? And how did we end up racing cars that resemble something we might see on the highway? That's what this chapter is all about, what NASCAR is and how NASCAR got to where it is today and a bit about what the future might look like.

American stock car racing actually predates NASCAR by several years. NASCAR actually came about due to good timing, a combination of racers searching for a way to make a living from racing, and the vision of a man who would create the sanctioning body millions of fans know today.

"I think it's important to learn about the history of our sport. We're all here for a reason and a lot of it is because of the pioneers of our sport. We owe them something."
—Driver Joey Logano, Martinsville Speedway, October 2016

The Beginning

NASCAR is actually an acronym. It stands for the National Association for Stock Car Auto Racing. Bill France Sr. formed this association in 1947, and it was incorporated early in 1948. Born in September of 1909, France found a fascination with the new automobile as a teenager. This technology was just taking hold in America. Bill took this fascination to a different level in his school years, sometimes skipping his high school classes to take the family sedan to a local track to turn laps, leaving time to get the Model T Ford back in time for his father to come home.[1]

Bill France Sr. held a number of jobs but eventually bought his own service station. Growing tired of the cold perhaps, the newly married France took his young wife, Anne, and his infant son to Daytona Beach, Florida, in 1934. At the time, the hard-packed sands of Daytona and its northern neighbor Ormond Beach provided the opportunity for thrill-seekers to push the limits of the new automobile. It had started in 1909, but by the 1930s, there were those trying to break the seemingly unreachable 300-mph barrier rocketing along the sandy beaches.

There are several legends surrounding the reasons Bill France Sr. decided to settle in Daytona—from the attraction to the speed record attempts, to a lack of money to go any further, to his car breaking down. Whatever the reason, Bill France Sr. worked a series of odd jobs before earning enough money to buy a service station on Main Street. During this time, France tried his hand at racing on dirt tracks in the local area.

France sensed something, and that "something" came at a fortuitous time. Just two years after France had moved to Daytona, those seeking speed records were finding the hard-packed sands to be prohibitive to their efforts thanks to tides and weather. They started looking west, to the Bonneville Salt Flats in Utah, a place less rutted and not susceptible to the ocean's tides or rain.

The Daytona Beach city leaders, desperate to make up the revenue lost by the abandonment of the speed record attempts, searched for a way to keep automotive competition, and the money that came with it, in their city. The leaders approached local dirt track racer Sig Haugdahl with the idea of a race along a 3.2-mile course that consisted of the paved road along A1A combined with the beach course formally used for speed record attempts. Haugdahl enlisted France, and the two secured a $5000 purse from the city. In 1936, The American Automobile Association (AAA) sanctioned the first race consisting of sedans that were street legal.

All did not go well, however.

The course had turns of mainly sand that saw many cars get stuck. In addition, there was very little that could be done to keep people without tickets from arriving early and staking out a spot. The full scheduled 78 laps never happened, due to problems with the track. The race was called three laps early; France finished fifth. The city estimated they lost over $20,000, making it the last beach automobile race the city would promote.

Undeterred, Haugdahl and France approached the local Elks Club the next year. They convinced the organization to help promote a Labor Day race on the same course. There was only a $100 purse, but the course, officiating, and promotion was improved. However, the Elks, like the city, lost money. After that second attempt, Haugdahl had had enough. That left any future promotion to Bill France, and Bill France alone.[2]

During this same period just up the road in Georgia, drivers working for a man named Raymond Parks were dominating most races across the South. Parks was a self-made millionaire from Dawsonville, Georgia. How he made that fortune, however, was somewhat dubious.

Moonshine, homemade liquor free from taxation, was popular in the South during Prohibition. Even after Prohibition, however, the South still had an appetite, and a there was still a market for moonshine.

Parks earned a decent living first delivering homemade liquor for others, then later making and selling his own moonshine.

Parks also owned a real estate business and a company that supplied jukeboxes, pinball machines, and other novelty machines to bars and restaurants. He spent time in jail when the law intruded on his moonshine efforts, once for a nine-month stretch during the mid-1930s.

Parks teamed with local mechanic Red Vogt and enlisted drivers such as Roy Hall and Lloyd Seay to ship his products to cities around Georgia, mainly Atlanta. With cars faster than those driven by local law enforcement thanks to the mechanical prowess of Vogt, and the driving skills of Hall and Seay, Parks could get his moonshine to the bars and restaurants in Atlanta with few issues.

On the weekends, when he was not running moonshine, the cars owned by Parks and driven by his skilled wheelmen could often be found at a dirt track somewhere in Georgia dominating the competition.

These loosely organized events drew not only moonshine runners, but working class people who would bring the cars they drove every day during the week, paint a number on the doors, and race. There was little, if any, money on the line, simply bragging rights and perhaps a cheap trophy.

It was into this world that France was introduced in 1938. Though he knew of Parks only by name, he was friends with Red Vogt. Vogt invited France to race at Lakewood Speedway near Atlanta. What France saw when he arrived would be a spark. Standing in the infield of the one-mile dirt track, he saw fans filling the grandstands—some estimates put the number near 6,000 people. Those not in the stands watched and cheered from the infield. France would finish that race behind Parks's drivers; however, he knew that the future was not only on the track, but also in the stands and the infield, where people paid to watch drivers compete against one another.

1940s

That visit to Lakewood gave France an idea. With no money of his own, France approached a local Daytona businessman with the idea of promoting the race on the beach. He was able to convince Charlie Reese, a local restaurateur, to put up a purse of $1,000. France promoted the race and recruited drivers. The 150-miler was by all accounts well attended. France finished second in the race, but more important, the event actually showed a small profit. With this momentum, France convinced Reese to continue backing the races and helped stage another successful event on Labor Day weekend in 1939. The following year three races were held in March, July, and September, with France winning the July race. In 1941, there were two races in March and another in July.

Meanwhile, France was still racing in the South and promoting races. He saw the potential of drivers such as Hall and Seay to become stars in a fledgling sport that was gaining popularity in the South, and selling tickets. France knew, though, that for the sport to gain real popularity, the image of a bunch of moonshine runners would have to go. It was then that Red Byron appeared on the scene. Byron had no moonshine roots and was raised in Colorado. He was a fearless driver on the track, and France saw him as the first non-moonshine-running superstar. His stable of superstar drivers attracted fans and sold tickets wherever they raced, and France had high hopes for the future. Those hopes would take a blow, however, when on September 2, 1941, Lloyd Seay, just twenty-one years old and a cousin of Parks, was gunned down in a family dispute. The loss of Seay hit France hard, not only because he considered Seay a friend, but also because he was a big draw at the track.

The second blow for France came as he was beginning preparations for racing in 1942. December 7, 1941, the day the Japanese attacked

Pearl Harbor and America entered World War II, changed the world forever, on so many levels, and even affected racing. There would be no racing on the beach road course again until 1946. During the war, Bill France worked at the Daytona Boat Works while his wife Anne ran the filling station. He and Anne also celebrated the birth of their second son, James, born on October 24, 1944.

It didn't take long for France to return to racing after the war. On December 2, 1945, Seminole Speedway, a quarter-mile dirt track located in Casselberry, Florida, just north of Orlando, held its first race. France promoted the event and finished second. Roy Hall, who won that first race, Red Byron, and Fireball Roberts were among the men who raced there and would later become NASCAR legends. Byron's story was even more remarkable considering he had overcome an injury sustained while flying in the Pacific Theater. His left leg was permanently damaged, and it was Red Vogt who modified the car Byron raced at Seminole. Byron won his first postwar race at Seminole in February 1946, beating Hall, France, and others, while driving a Raymond Parks-owned, Red Vogt-modified car.

During those years immediately after the war, France not only promoted the events at Seminole Speedway, but continued racing himself all throughout the South, bringing along with him some superstar drivers who drew in the crowds. At this time in history, dirt tracks were scattered all around the region. Those who raced at these tracks, including France and Byron, were not all moonshine runners, contrary to what NASCAR lore portrays. While many of the racers did in fact spend their nights hauling moonshine in cars, the drivers themselves modified, just as many were simply blue-collar folks looking for a little excitement and bragging rights.

Several sanctioning bodies controlled these races across the South. The rules were loose, the purses inconsistent, and unscrupulous promoters and track operators sometimes even took the money collected at the gate and disappeared, leaving the competitors with empty pockets. In addition, none of these sanctioning bodies crowned a national champion.

France saw a need for organization, and on December 12, 1947, he held a meeting at the Streamline Hotel on A1A in Daytona Beach. He called together track and car owners including Parks; drivers including Byron, Hall, and Roberts; and mechanics including Vogt. There were discussions about the future of the fledging sport of stock car racing, and it was then that France first proposed the National Association for Stock Car Auto Racing.

The idea was to bring together the sport under the umbrella of a single sanctioning body, one that had consistent rules, rules that ensured that all drivers had an equal chance to win a race. That was something France knew would attract fans, a level playing field. Prize money would also be consistent and allow those in the sport to have an opportunity to make a full-time living racing stock cars. Finally, NASCAR could be the organization that would crown a national champion. Legend has it that France developed the first points system and wrote it down on a bar napkin during these first meetings.

NASCAR was incorporated on February 21, 1948. Only days before that incorporation, NASCAR sanctioned its first race on the Daytona Beach/Road course. An estimated 14,000 fans attended the race, which Red Byron won.[3]

The original plan by France was to have three classes: Strictly Stock, Modified, and Roadsters. It didn't take long, though, for those running NASCAR to see that southern fans had little interest in the Roadster, a car normally associated with racing in the North.

The Strictly Stock class, which we now know as the top premier Cup series, really didn't take off that first year of 1948. The American automakers had been in the process of converting from wartime manufacturing back to building cars. Thus in 1948, they couldn't keep up with consumer demand in the postwar world. It was left to the Modified division (which today is known as the Whelen Modified Tour) to carry the NASCAR banner that first full year. There were 52 NASCAR-sanctioned races for the Modified division in 1948, with Red Byron crowned as the first NASCAR National Champion.

The Strictly Stock division began to get its wings in 1949, but just barely. The very first NASCAR-sanctioned Strictly Stock race was held in Charlotte, North Carolina, at the Fairgrounds Speedway on June 19. Jim Roper won the race, Bob Flock won the first pole, and Sara Christian, who finished 14th, was credited as the first woman to race in NASCAR's premier division.

There would only be eight races in that first season (today there are 36). From Charlotte to the beach/road course in Daytona to New York, Pennsylvania, and North Carolina, the first season consisted of all-dirt tracks with the lone exception of Daytona, which was only partially paved. Red Byron won the first NASCAR Strictly Stock championship on October 16, 1949, driving for owner Raymond Parks. Winning cars ranged from 1949 model Lincolns, Oldsmobiles, and even a Plymouth driven by future superstar Lee Petty, who won his first NASCAR race at Heidelberg Raceway in Pittsburgh on October 2.

1950s

The schedule was expanded to 19 races for 1950 and opened in Daytona Beach on February 5. France also changed the Strictly Stock moniker to Grand National, the name the premier series would be known as until 1971. The series would eventually race a high of 62 races in 1964 and would pare down to the 36 races we have today in 2001.

Nineteen-fifty was a pivotal year for the growing sport. France knew he needed a big draw, something to compete with the AAA-sanctioned Indianapolis 500. The Indianapolis Motor Speedway was already forty-one years old, and the Indianapolis 500, a race featuring open-wheel roadsters, was the most prestigious auto racing event in America. France knew he needed some sort of event that could compete with Indy. It was an old friend, Harold Brasington, who came to the rescue.

Brasington had raced with France, and the two were friends. In the Fall of 1949, Brasington purchased a 70-acre farm near his hometown of Darlington, South Carolina. His dream was a superspeedway, and soon he was constructing a 1.25-mile, high-banked, egg-shaped speed-way on what was once cotton and peanut fields. The new speedway was egg-shaped instead of a true oval because Brasington kept a promise to the original landowner that he wouldn't disturb a minnow pond on the property's western fringe.[4] This meant that while the eastern end of the new track could be as wide and sweeping as needed, the western end had to be a bit narrower. NASCAR still races at Darlington to this day, and the speedway, with its opposite turns, remains one of the most challenging on the circuit.

Back in 1949, Brasington did something that stunned the racing world: he paved his brand new superspeedway with asphalt, making it the largest fully paved track in the South. The track opened in 1950 and held the inaugural Southern 500 on Labor Day. The Southern 500 marked NASCAR's longest race (the race is actually 501.322 miles long) and took two weeks to qualify a field of 75 cars and a full six hours to complete. Nine thousand fans sold out the grandstands, and the overflow, estimated to be over 15,000, watched from the infield.

Darlington marked an era of growth for NASCAR. Thanks in large part to the Southern 500, the sport of NASCAR began to take hold in the South. Drivers with names like Fireball Roberts, Buck Baker, Herb Thomas, the Flock brothers (Bob, Fonty, and Tim), and Lee Petty became stars. More paved tracks and even a couple of road courses were added to the schedule, and soon NASCAR was becoming a highly organized way of life—a far cry from the unorganized, unsanctioned events from the recent past.

Bill France Sr. still had an eye for the future. In 1953, France founded International Speedway Corporation (ISC). While not officially con-nected to NASCAR, the company was formed to build and manage the tracks at which NASCAR holds races. Today, ISC is a multibillion-dollar company that owns and manages 13 tracks all across the United States. In 1953, however, France had a dream, a dream that could only come about through ISC. France wanted a superspeedway the likes of which the world had never seen. He wanted a track to replace the

Beach/Road course, one with high banks to allow for fast speeds and better viewing for fans. He wanted a track that rivaled the Indianapolis Motor Speedway. So in 1953, France got to work on the design, and in the ensuing years he approached the city of Daytona Beach, got an agreement to lease 450 acres five miles from the Beach/Road course, and began to raise the funds. On November 25, 1957, they broke ground. The 2.5-mile Daytona International Speedway was unlike anything anyone had ever seen or on which any driver had ever raced. The track still opens the NASCAR season today, and millions of fans have been thrilled by the races in the ensuing years. But in 1958, the future was uncertain for the speedway. France had run out of money; he mortgaged his house, begged investors, and took a chance by selling tickets to a race that only France had the vision and confidence to know would happen.

The first stock cars tested at the new track in early February 1959, and on February 22, 1959, nearly 42,000 paying fans watched the first-ever Daytona 500. That inaugural event delivered the kind of racing France hoped for—stock cars racing faster than anywhere else in America. Lee Petty beat Johnny Beauchamp across the line by inches, but that wasn't official on the day of the race. On that day, Beauchamp was declared the winner. Petty insisted that he had beaten Beauchamp, and it took photos and three full days to back Petty's claim up. Legend has it that the wait was made a bit longer as France milked the controversy in a masterful PR move that kept the story on the front pages of newspapers until France declared Petty the victor. That first Daytona 500 set the stage, and more and more people began to notice the sport of big-time stock car auto racing.

1960s

The 1960s began with the first national television exposure for NASCAR. The CBS Sports magazine show "Sports Spectacular" aired the 100-mile qualifying races from Daytona International Speedway on

February 12, marking the first time NASCAR races were exposed to a national audience. The ensuing Daytona 500 was also shown, but only a part of it and on a taped delay. This would be the story for NASCAR and television throughout the 1960s. Flag-to-flag live coverage of an entire race wouldn't happen until 1979.

In 1960, speedways in Atlanta and Charlotte opened. Unlike today, the Grand National schedule wasn't a set one that had the same competitors show up at the same track each weekend. Nineteen-sixty saw a total of 44 races, some held on the same weekend or only days apart. Most races were still held on dirt, but during the decade, as more speedways opened, more of the older dirt tracks were paved. In turn, the dirt tracks began to fall off the schedule.

The 1960s also ushered in new stars. There were names like Junior Johnson, a former moonshine runner who still occasionally hauled shine and even missed some races after getting caught and having to serve jail time. There were Joe Weatherly, Ned Jarrett (nicknamed Gentleman Ned), and Richard Petty—the son of Lee Petty, former champion and winner of the first Daytona 500. Richard Petty arguably made the biggest splash in the sport during the 1960s. Richard competed in his first race in 1958 and was Rookie of the Year in 1959. That year, Richard won his first NASCAR race, at Lakewood. However, his father, Lee, protested the results and was ultimately awarded the win.

It was the marriage of Richard Petty, Plymouth, and its new Hemi engine in 1964, however, that set the NASCAR world on fire. By 1960, the auto manufacturers understood the power of marketing that NASCAR could provide. It was during this era that the term "Win on Sunday, buy on Monday" was first used. The big three automakers backed the top drivers in NASCAR knowing that the potential buyers in the stands could be influenced by their heroes on the track.

The marriage of Richard Petty and the Plymouth Hemi produced immediate results. In 1964, Petty led all but 16 laps en route to his first Daytona 500 win (his first of a record seven Daytona 500 wins, a feat that has not been equaled to this day). He won nine races in 1964 and the first of his seven Cup titles, a record that only Dale Earnhardt Sr. and Jimmie Johnson would tie.

Petty's popularity helped NASCAR's growth, but in 1964, Petty's and Plymouth's success raised the ire of Bill France Sr., who felt that having one driver and manufacturer dominate the races could alienate fans of other drivers and manufacturers. That year, Hemi power won 26 of the 62 races on the schedule. France banned the Hemi engine for the 1965 season. France's reasoning was that the Hemi was not readily available to the public in production models, not "stock" enough, which was still the foundation of NASCAR—cars racing on the track that closely resembled those driven on the street and available to the general public.

Plymouth elected to boycott NASCAR, pulling its factory support from its NASCAR teams. Petty followed Plymouth's lead, and suddenly one of NASCAR's biggest early stars left the stock car tracks to go drag racing.

The 1965 season opened without a Petty on a NASCAR track. Other stars who raced under the Plymouth banner also sat out part of the 1965 season, including Bobby Issac, David Pearson, and Paul Goldsmith. Ticket sales suffered, and Ford branded cars won the first 34 races of the season. France got a wake-up call. The momentum the sport's popularity had enjoyed was waning. Without Plymouth, Ford was the dominant manufacturer, and without the stars of Plymouth, seats were empty in the stands. France lifted the Hemi ban halfway through the 1965 season, ruling that the Hemi could be used only in the full-sized cars raced on the smaller tracks, but not the midsized models raced on superspeedways. This allowed France to save face while allowing some of the sport's biggest draws at the time to return. The Plymouth drivers returned, and Petty won three races.

France may have thought his manufacturer issues were done as the 1965 season ended; however, they were far from over. Ford had enjoyed domination in 1965. Without the competition of the Hemi, Fords were the class of the field each week. Once France allowed the Hemis back, though, Ford knew it needed an answer. That answer came in the form of the 427 cubic-inch single overhead cam engine (SOHC).

Ford had already tried to introduce the 427 SOHC but had been rebuffed. In an earlier attempt to compete with the Hemi, Ford had sought approval for the SOHC prior to the 1964 Daytona 500. But France had denied the approval for the 427 SOHC, which was considered exotic and not readily available in passenger cars.

Prior to the start of the 1966 season, though, Ford tried again. Instead of approaching NASCAR and France to seek approval, Ford announced late in 1965 that the 427 SOHC would be its primary engine for 1966. The announcement came on the same day France was leading a media event at the Chrysler plant showing off a line of production Hemi engines that would be sold in passenger cars.

France knew he was facing a big threat to his, and NASCAR's, authority. France sought an alliance with the other major American motorsports sanctioning body at the time, the United States Auto Club (USAC), which dealt primarily with Indy cars and open wheel racers in the Midwest. Together, NASCAR and USAC issued a statement declaring that since the Ford 427 SOHC engine wasn't available for passenger cars, the engine would be banned for racing. Ford responded by announcing they would not be competing in 1966. Remembering the debacle caused by the lack of Hemis and Chrysler stars in 1965, France called a meeting with Ford officials. On Christmas Day 1965, NASCAR and Ford announced that Ford would be racing in 1966. The separate announcements came after NASCAR assured Ford that as soon as production cars with the 427 engines were rolling off the line, they would be allowed in NASCAR.

Ford rushed to put their 427 SOHC engines into production, or least make it appear so. For the first two races of the 1966 season, at Riverside and Daytona, Ford was outclassed and their cars uncompetitive. Ford detuned the powerful engines and somehow put them in production cars. They even had pictures showing new cars with the

427 badge at several new car dealerships to prove they were being sold. According to racefansforever.org, however, the pictures were actually of the very same car repainted and moved to different Ford dealerships around the country.

France knew the Ford PR campaign was just a smoke screen. He also knew, however, that if he were to continue the Ford ban, the manufacturer could institute a boycott, and France still had the fresh memories of the Chrysler boycott. France decided to leave the decision up to someone else, perhaps to save face. He approached the Fédération Internationale de l'Automobile, or as it is commonly known, the FIA. The FIA is an association established as the Association Internationale des Automobile Clubs Reconnus in 1904 to represent the interests of motoring organizations and automobile users. The FIA came up with a compromise—Ford could race the 427 SOHC engine but would have a weight penalty of just over 400 pounds. The FIA did allow Ford to have a second carburetor on the 427, though, that would help it compete with the Hemi.

Ford was not pleased.

Over the course of the five days after the decision, there were three short track events. Ford announced that it would not enter any factory-backed Ford cars in those races. It was a somewhat strange decision, since teams would not be running the new engine in those events anyway. Prior to the next speedway event on April 15, Ford said they were pulling their teams and instituting a NASCAR boycott. They further warned any team running Ford equipment would have their factory contracts and backing terminated if they entered any NASCAR events as an independent or drove another car. At the time, those drivers competing for Ford included Bobby Isaac, Fred Lorenzen, and reigning Grand National champion Ned Jarrett.

France called a meeting with his track owners soon after. At the meeting, the owners decided to stand with France and put up a united front against Ford. A PR campaign was launched to show fans that Ford wasn't acting in their best interest. This brilliant move didn't help

Ford sell any cars. On the track, some independent drivers were still racing Fords and winning at the short tracks. A further PR disaster for Ford came when one independent driver, Tom Pistone, tried to buy a 427 SOHC engine. Despite the fact that Ford claimed the engines were in production for nearly six months, Pistone was unable to find a single one.

It was Ford's own factory drivers, however, that helped end the boycott. Not content to sit on the sidelines, many drivers gave up their lucrative factory contracts and went back driving for other owners and manufacturers. The last driver was Lorenzen, who still abided by the Ford boycott. It was Lorenzen who helped end the boycott, aided in large part by what could be considered another brilliant PR move by France. Mr. France could gauge that the tide was turning against Ford and decided to make a few concessions. After all, Lorenzen was one of the most popular drivers in the sport, and fans filled the stands to see him race. At Atlanta on August 7, Lorenzen returned to the track for the first time since the boycott with a Ford prepared by Junior Johnson. That Ford, however, looked like nothing that had raced before. The Ford Galaxie looked almost nothing like its production counterpart. It was lowered significantly with a chopped roof and a sloping front windshield. It was so illegal that other competitors called it the "Yellow Banana." Despite this, though, the car passed NASCAR inspection. This outraged everyone else in the garage. The "Yellow Banana," with Lorenzen behind the wheel, led 24 laps that day but crashed out after lap 139 and never finished the race. The uproar didn't end there, though, as track promoters vowed that such a car would never race on their track. France admitted that the rules at Atlanta had been a bit "gray" and assured everyone that he had instructed Johnson to retire the Yellow Banana. But the damage for Ford in the aftermath was immeasurable. Despite the fact that they had nothing to do with the Lorenzen car, the perception was that they were somehow behind it. Did France conspire with Johnson? We may never know, but the results for France were more butts in the seats at Atlanta after the incident to see Lorenzen race and mud on the face of Ford, which looked like a cheater in the eyes of fans.

On a side note for that Atlanta race, famous mechanic Smokey Yunick, known for pushing the envelope of NASCAR's rules, prepared a Chevrolet Chevelle for that same race that was almost as illegal as the Yellow Banana. The car was 7/8 scale, and perhaps taking advantage of the lax inspection for that race, Yunick saw it pass inspection with no issues. Curtis Turner raced the car but was forced to retire early with engine issues.

That August 1966 race was a turning point of sorts for the growing sport of NASCAR. France had faced off with the major manufacturers, and it was the manufacturers who had blinked this time. France asserted his dominance and control of the sport and let everyone know just who was in charge. That authority would be further tested and reasserted as the decade wound down.

The 1960s for NASCAR represented the era when the sport really began to grow. With factory backing for teams, organized pit crews, star drivers, and more sponsorship dollars, NASCAR was becoming more professionalized. People could now make a living racing, something that France had always wanted.

While there were many star drivers during the 1960s, perhaps none was bigger than Richard Petty. The young Petty had an easygoing way that charmed all who met him and gained him ever-increasing legions of fans. On the track, Richard Petty was a talented driver and fierce competitor. During the decade, Petty won 101 races and earned two of his seven titles.

With Richard Petty leading a parade of star drivers on the track, fans were filling the stands, and the garages were filled with people who earned a living in the sport.

France wasn't fully satisfied, however. After all, he had had a vision for a track similar to Daytona, but one that was bigger, longer, faster, and could showcase the speeds at which NASCAR raced. His first idea for a location for this new track was in North Carolina at the site of the Occoneechee Speedway, which became the Orange Speedway in 1954. France had built the 0.9-mile dirt track on the site of an old horse track in 1947. When France wanted to build his new super-speedway on the site, the government of Orange County rebuffed his efforts. France began to look elsewhere and found a spot in the Deep South along Interstate 20 between Atlanta, Georgia, and Birmingham, Alabama. He found the site of the former Anniston Air Force Base, purchased the land, and on May 23, 1968, ground was broken on the new Alabama International Motor Speedway (in a nod to the government of Orange County, France closed Orange Speedway shortly after the groundbreaking in Alabama).

While a tri-oval like Daytona, the track was 2.66 miles, as opposed to the 2.5 miles of Daytona. It was not only longer, but also wider and with steeper turns (33 degrees rather than 31) than Daytona. France added one last touch to his new facility, partially due to the lessons learned from Daytona. The start-finish line was moved nearer to turn 1 rather than mirroring the one at Daytona, which was at the middle of the grandstands. France reckoned that, this way, he could sell more tickets—not only that area, but in total, too, since fans could watch drivers make one final move to win. The slingshot move, when the rear car uses the draft of air from a front car to get momentum for a pass, had been in use at Daytona. France knew that the extra 1250 feet could give drivers one last chance for a win. He was correct, and since its opening, many of the finishes at the track first named the Alabama International Motor Speedway have been memorable ones. The name of the track changed in 1989, and today Talladega Superspeedway remains one of NASCAR's biggest venues with races that are loved by fans and feared by competitors.

The first race at Talladega almost never happened, though.

It took sixteen months to finish the new speedway—a bit longer than anticipated, thanks in part to Hurricane Camille, a Category 5 storm that roared ashore along the Mississippi on August 17, 1969, less than a month before the first scheduled race. While the storm caused catastrophic damage and over 200 deaths, the speedway crew worked overtime to repair the damage to the new speedway and prepare for the first race. They were able to, and Mr. France probably thought his troubles were over. But it turned out they were just beginning.

The first Grand National race was scheduled for September 14, 1969. In the days leading up to that first race, trouble began to brew. The track was the widest, longest, and fastest that any stock car had ever been on. During testing, the tires used at the time had issues keeping up with speeds nearing 200 miles per hour. Tires shred and blew, and while no one was injured, considering the lack of safety advancements at the time, that was due to luck more than anything. It wasn't the fault of the tire manufacturers; they simply didn't have the technology to keep up with the increased speeds and new pavement.

Drivers, those once thought fearless speed demons, were concerned.

The opening of Talladega came just a week after drivers had met in Detroit and tried to form a union. Led by NASCAR's biggest star at the time, Richard Petty, the Professional Drivers Association was formed. The PDA included David Pearson, Cale Yarborough, Buddy Baker, and Bobby Allison. The goal of the PDA was to increase safety, raise incomes, and lobby for guarantees. In 1969, some of the Grand National races paid only around $1000 to the winner.

This wasn't the first time in NASCAR history that drivers had tried to organize. In 1961, drivers Curtis Turner and Tim Flock partnered with the Teamsters in an attempt to unionize the drivers. Mr. France would have none of it, and he quickly squashed the attempt. Both Turner and Flock were banned "for life" from NASCAR. The bans were lifted four years later, and while Turner would return to a NASCAR track, Flock never raced in NASCAR again.

The 1969 formation of the PDA came at a fortuitous time, because Talladega presented the first real test of the PDA. Charlie Glotzbach won the pole with a speed of 196.386 mph. There was a famous showdown, however, that put the race in jeopardy. On the Saturday prior to the race, PDA drivers, led by President Richard Petty, confronted

France in the garage. They pointed to the high speeds, tire failures, and unsafe conditions. France was asked to postpone the race, and he of course refused. According to the *Daily Press*, France told the drivers if they didn't like driving fast, they should just pack up and leave, and that's what most of them did.

Knowing there were fans with tickets, France put together a field of 36 drivers, most from other NASCAR series, and staged the race. The Professional Drivers Association was the biggest loser. A week after Talladega, all the drivers who had boycotted the event were racing again. The PDA was never heard from again, and France proved that the sport of NASCAR was bigger than its drivers were and always would be.

The PDA was a wake-up call for NASCAR and France, however. As the decade of the 1960s ended, an era of better treatment for NASCAR's stars was dawning, an era that also saw a changing of the guard in NASCAR and saw it reach heights its founder could never have imagined.

1970s

The early 1970s marked big changes in NASCAR. Moving forward, all the top series races were on pavement, at facilities that were becoming more and more modernized and fan-friendly. International Speedway Corporation was growing. In 1970, ISC founded the Motor Racing Network, a radio broadcast entity devoted to flag-to-flag coverage of Grand National races. MRN had its first broadcast, the Daytona 500, on February 22, 1970. On September 30, 1970, the final NASCAR Grand National race on dirt was held at the State Fairgrounds in Raleigh North Carolina.

Behind the scenes, NASCAR continued to work on its growth. And for the first time in its history, that growth was led by someone other than the man known as William "Big Bill" France. Bill France Jr., the son of Big Bill France, was born in Washington D.C. a year before his parents moved to Daytona Beach. The younger France had started in the sport at an early age, with the focus on promotions. He worked at the old Daytona Beach course as a youth, selling concessions and parking cars, among other behind-the-scenes duties.

While France Jr. was working his way up the NASCAR ladder, his younger brother Jim started working for his father at ISC in 1959, when he was just 14. Like his older brother, Jim would park cars, sell tickets, and perform all manner of mundane tasks at the tracks. He would serve in the Army in Vietnam in 1969–1970 and would eventually ascend to become president of ISC in 1987. Today, he serves as ISC chairman.

France Jr.'s real involvement in NASCAR began with the construction of Daytona International Speedway. During construction, France Jr. worked seven days a week at the speedway, sometimes even operating the heavy equipment, and even using mules to pull stumps out.

In 1966, the elder France appointed his oldest son as vice president of NASCAR. It was in that capacity that France Jr. began to show his knack for promotion that would serve the growing sport. In 1971, France Jr. negotiated the deal that was perhaps the biggest thing NASCAR had seen up to that point.

That year, the federal ban on cigarette advertising, part of the Public Health Cigarette Smoking Act of 1969, went into effect. Major tobacco companies were looking for ways to use their advertising dollars once used for television advertising. Junior Johnson approached RJ Reynolds looking for sponsorship of his team and then approached France Jr. for help. Bill France Jr. saw a bigger opportunity, however. In 1971, the announcement was made that, thanks to a partnership with R.J Reynolds Tobacco Company, the NASCAR Grand National Series would be known as the NASCAR Winston Cup Grand National Division and the series' yearly national championship title would be the NASCAR Winston Cup.

That wasn't the only big change for the early 1970s. In addition to a new title sponsor, the schedule was pared down from 48 to 31 races a year. Then on January 10, 1972, Big Bill France announced that he was turning over control of NASCAR to his son Bill France Jr., who became only the second NASCAR president in the sport's history.

That year, 1972, is cited by many as the start of NASCAR's "Modern Era." During the first years of the Modern Era, the yearly schedule became firmer, races on ovals shorter than 250 miles were dropped, and in 1974, a new points system would be introduced. France Jr. had tasked series public relations representative Bob Latford, who had a talent for numbers, with developing a points system that could be used during an entire season, regardless of a race length, to decide a national champion. Winston would remain the title sponsor through 2003, and the Latford points system would be used until 2011.

France Jr. secured the title sponsor and looked to get more exposure for the sport nationally, primarily with television. ABC had tried broadcasting a full race, the 1971 200-lap Grand National race at Greenville-Pickens Speedway, but the race itself wasn't as exciting as most, and the network returned to showing only taped edited highlights of races.

The 1976 Daytona 500 provided one of the first memorable TV moments in NASCAR history. David Pearson and Richard Petty were shown battling for the win on the final lap. The two were side by side exiting turn 4 but crashed nearing the finish line. The mangled machines came to rest in the infield, with Pearson moving his damaged Mercury first and nursing it to the checkered flag moments later.

Not content with only highlights, France Jr. worked behind the scenes to secure a TV contract for the sport. In 1978, he finally got the attention of network executives, specifically CBS Network's president, Neal Pilson. CBS agreed to take a chance on broadcasting an entire race live, and on February 18, 1979, CBS broadcast the Daytona 500 live flag-to-flag. It turned out to be a watershed moment in the history of NASCAR, thanks to a snowstorm and a dramatic finish capped off with a postrace fight.

It didn't start out that way, however.

The morning of race day dawned under overcast skies with persistent rain. Bill France Jr. almost saw the dream for which he had worked so hard slip from his grasp, thanks to Mother Nature. CBS producers told France that they were nearing the point where alternative programming would have to be considered. France told the producers and executives in attendance in no uncertain terms that not only would there be a race, but it would start on time. Shortly after the rain stopped, the sun broke through the clouds. But there was still a problem. The only way to dry a track as large as Daytona then was with time and trucks that dragged old tires around the track. France needed to start the race on time for TV, and for the first time in NASCAR history, a race started under green-yellow flag conditions (meaning laps would count, but no passes could happen and drivers had to follow the pace car). Nothing dried a track faster than a field of cars rolling around, and the first 30 or so laps of the 1979 Daytona were run under yellow flag conditions as the track dried.

While Daytona was dealing with rain, farther up the East Coast, the majority of residents were dealing with a monumental snowstorm. The storm kept many inside, and with limited TV channels in those days, countless snowbound residents tuned in to see just what this NASCAR racing was all about.

The 1979 Daytona 500 was a typical exciting affair with 36 lead changes. The final laps featured the biggest stars of NASCAR at the time dueling for the win. Donnie Allison, who had led a race-high 93 laps, was leading the field exiting turn 2. He was closely pursued by the tenacious Cale Yarborough. As the two rocketed toward turn 3, they got together and spun to a stop in the grass near the bottom of turn 3. The crash opened the door for Richard Petty, who charged to victory. At first, the TV cameras were focused on Petty as he celebrated his surprise win, but moments after the checkered flag was waved, the attention fell on the area where Allison and Yarborough had stopped. Donnie's brother, Bobby, fearing for his brother's safety, had stopped his race car at the scene. Soon Bobby and Yarborough were seen throwing helmets and punches at each other.[5] It was pure TV gold.

The action on the track and off, coupled with the snowstorm, led to huge ratings that left TV executives smiling and wanting more. Bill France Jr. was more than willing to give them just that.

The 1979 Daytona 500 was also memorable for another reason. A young rookie made his Daytona 500 debut that year and led 10 laps, finishing eighth. That young driver was named Dale Earnhardt, a man who would become a legend and a driver who would lose his life in the Daytona 500 in 2001.

1980s

The 1970s ended with Richard Petty winning his seventh, and what turned out to be his final, Cup championship in 1979. Petty represented a group of mature drivers who were at their prime, while the decade of the 1980s represented the rise of NASCAR's youth, which would in turn attract new fans. It would also bring even more national attention to the sport, as it continued its rise from a regional sport.

Along with Earnhardt, drivers with names such a Rusty Wallace, Darrell Waltrip, and Bill Elliott challenged the established names of Petty, Yarborough, and Allison. Earnhardt made perhaps the strongest statement of these young drivers as the decade started. In 1980, the twenty-nine-year-old driver, whose father Ralph had raced in NAS-CAR from 1956 to 1966 and who died in 1973, won five races and his first of seven titles in his first full year in NASCAR. It was then Darrell Waltrip's turn. The thirty-four-year-old driver from Owensboro, Kentucky, won 12 races in 1981, more than enough wins to secure him his first title. Waltrip would mirror his win record the following year and add another Cup championship to his résumé.

Bobby Allison would be the last of the old guard to win a title in 1983. By then, however, the young guns were starting to lead the way.

In fact, it was Waltrip who became a bold, brash, and colorful figure who never hesitated to speak his mind, who according to legend once said the "old guys better watch out because there's some new kids in town."

In 1982, NASCAR brought together the regional racing series known as the Late Model Sportsman Division into a new series. Rising costs made it harder and harder for regional racers to compete on a weekly basis at bigger races with bigger purses. This new series was known as the NASCAR Budweiser Late Model Sportsman Series. Many of the races were held at historic venues like Hickory in North Carolina and South Boston, Virginia. Many of the events, however, were held as companion events to Winston Cup races at Daytona, Richmond, and Bristol. This series was rebranded in 1984 as the Busch Grand National Series and the NASCAR Busch Series Grand National Division the following year. It would become the NASCAR Busch Series in 2004 and remain that until 2008, when Nationwide Insurance took over as the title sponsor and the series became the NASCAR Nationwide Series. In 2014, Nationwide left the series, and Xfinity, a subsidiary of Comcast Cable, became the title sponsor.

The involvement of Anheuser-Busch in the Late Model Sportsman Series shows the growing attraction of national companies to the sport of NASCAR. In the 1980s, many companies looked to NASCAR to increase brand awareness.

Drivers like Waltrip could be seen sporting the colors of Tide detergent or Mountain Dew soft drinks. Richard Petty was the first driver to carry a big-name sponsor, though. In 1972, he was paid a quarter of a million dollars to sport the STP logo on his car. The company wanted the car painted red to match the colors of the branding of its automotive additives. Lee Petty, however, who had painted the cars blue, was less than enthusiastic about this. But not wanting to turn away money, the two sides came to a compromise, which became a red-and-blue paint scheme that has become synonymous with Richard Petty to this very day.

It was not until the 1980s, though, that national advertisers started to see the value of branding NASCAR race cars, thanks to an ever-increasing number of fans not only at the track, but more and more on television, as well. By 1989, every race in NASCAR's top Cup series was being broadcast. That growth was due in large part to a young upstart cable network, the Entertainment and Sports Programming Network, better known as ESPN.

ESPN had been on the air for only two years when it broadcast its first race from the North Carolina Motor Speedway in 1981. During this time, networks and the tracks negotiated individual rights deals leading to a patchwork of networks showing races, but by 1985 all NASCAR races were being shown either live or tape delayed nationally on CBS, ABC, ESPN, TNN, TBS, and syndication services such as Mizlou.

Although fans almost needed a separate playbook to find out where to tune in, what they saw on the track was a group of NASCAR stars who represented some of the best drivers in the history of the sport, both young and old.

Richard Petty was still a force to be reckoned with. Fans sporting red and blue filled the stands each week to cheer on the driver who was being called the "King" of NASCAR. On July 4, 1984, the King delivered one of the most memorable wins in sports history. Petty won the Firecracker 400 at Daytona International Speedway with President Ronald Reagan in attendance. It was Petty's 200th career win. Although Petty would race until 1992, that win was his last. It set a race winning mark that no one in NASCAR has come close to equaling, and probably never will.

While older fans cheered for Petty, many new fans began to follow the young Dale Earnhardt. In his first full year in the Cup series, 1980 (he ran 27 of 31 races in 1979 and won Rookie of the Year), Earnhardt won five races and his first of seven Cup titles. To this day, Earnhardt remains the only driver to win Rookie of the Year honors and a Cup title the following year. His seven titles, including three in the 1980s, tied him with Petty and later Jimmie Johnson. Those three drivers remain as the only NASCAR champions with seven titles. But unlike Petty, who was seen as a down-home, smiling good old boy, Earnhardt had a different aura. Earnhardt was gritty and not afraid to speak his mind. A high school dropout, Earnhardt carried a blue-collar persona that endeared him to the southern NASCAR fans, many of whom worked in farm fields, factories, and other blue-collar professions.

The 1980s were the years Earnhardt built his legend. In the years immediately following his first title, Earnhardt struggled. In 1982, he failed to finish 18 races and suffered a broken kneecap during a crash at Pocono. It wasn't until his move to Richard Childress Racing in 1984 that the legend of Dale Earnhardt really began to emerge.

Richard Childress had raced in NASCAR during the late 1960s until 1981. Childress's first Cup start came in the fabled September 1969 race at Talladega. Childress was one of the non-Grand National drivers tapped by Big Bill France as a replacement driver when the Grand National regulars boycotted the race. Childress finished that race 23rd, but with more money than he had ever made racing in his life. With that money, Childress built his own race shop in Welcome, North Carolina, about an hour north of Charlotte and just south of his hometown of Winston-Salem.

After a brief but unremarkable pairing in 1981, Earnhardt returned to Richard Childress Racing in 1984 and never looked back.

Earnhardt won his second title, and his first with Childress, in 1986. In 1987, Earnhardt won 11 races and his third title. That same year, Earnhardt earned the nickname "The Intimidator" after he beat Bill Elliott in the non-points race known as "The Winston" at Charlotte Motor Speedway. Among the many highlights from his career was the famous "Pass in the Grass" that occurred during that same race, which today is known as the NASCAR All-Star Race. While racing for the lead along the frontstretch, Elliott forced Earnhardt down the track. Earnhardt's car briefly went off into the infield grass, but the talented driver was able to keep control, and six laps later he scored the win.

Bill Elliott, who lost that race, was also building a legend of his own. The Georgia native, hailing from the stomping grounds of Raymond Parks, Dawsonville, had a personality that mirrored that of Richard Petty. As the aging Petty became less competitive as the 1980s wore on, it was Elliott who many Petty fans began to cheer for. Elliott won his first race in 1983 during his first full season. It was in 1985, though, that Elliott began to make his own legend in NASCAR. That year, series title sponsor Winston offered a $1 million bonus to any driver who could win three of the four races considered "Crown Jewels": The Daytona 500, The Winston 500 at Talladega, The Coca-Cola 600 at Charlotte, and the Southern 500 at Darlington.

Elliott won the season-opening Daytona 500 in 1985 and followed that up with a win at Talladega in May in dramatic fashion, overcoming a two-lap deficit. Elliott would go on to win the Southern 500 at Darlington in September, becoming the first (and as it turned out one of only two drivers) in NASCAR history to win the Winston Million (Jeff Gordon would mirror Elliott's wins in 1997 to win the $1 million bonus in its final year). Elliott gained national notoriety for the accomplishment, and it earned him a spot on the cover of *Sports Illustrated*. Elliott would go on to win the 1988 Cup title, 44 races, and NASCAR's Most Popular Driver title a record 16 times.

While Elliott was winning over fans with his easy style and boyish charm, Earnhardt was becoming a polarizing figure behind the wheel of his iconic number 3 Chevrolet.

On the track, Earnhardt was an aggressive no-nonsense driver not afraid to use the bumper of his race car to win races. In 1986, during the 400-mile race at Richmond in February, Earnhardt battled Darrell Waltrip for the win. Waltrip hunted Earnhardt nearly the entire race and finally passed Earnhardt with three laps to go. Entering turn 3 on the final lap, Earnhardt got to the bumper of leader Waltrip and tried to spin him out. The move backfired, though. Waltrip did spin, but Earnhardt was collected, as well. The incident didn't prove popular with Waltrip, many of the other competitors, and some fans. NASCAR didn't approve, either, and slapped Earnhardt with a $5,000 fine and a one-year probation, the latter of which was lifted soon after.

Off the track, Earnhardt was a man of few words, but with an innate business sense. Earnhardt had been married three times by the time he signed with Childress. His early life and career saw him barely making enough to support his growing family. Perhaps his aggressive racing style was motivated by winning more money, but whatever the reasons, Earnhardt sensed his popularity was increasing. With t-shirts and other items bearing the words "The Intimidator" selling to his ever-increasing fan base, Earnhardt created his own marketing company, Sports Image Inc., and soon "The Intimidator" was synonymous with Dale Earnhardt.

The Earnhardt move to form his own marketing company represented the growing involvement of corporate America in NASCAR. In the 1980s, not only were larger companies looking to advertise in the sport, but star drivers were beginning to realize that they were becoming brands themselves. The days of drivers speaking exactly what was on their mind were disappearing. As the checks from sponsors grew, they demanded more from their drivers. Drivers in turn started to become more politically correct and interlaced their conversations with the growing media corps with mentions of their sponsor.

Today, that practice is fully ingrained in NASCAR. Rare is the time when you will see a driver quote that doesn't mention a sponsor. That all started in the 1980s, the period when corporate America found the power of NASCAR.

That power was growing as the 1980s came to a close. Few could predict, however, that the 1990s would mark a meteoric rise in the popularity of NASCAR—something that founder Big Bill France himself could never have imagined back in 1947.

1990s

With every race broadcast on national TV as the 1990s began, NASCAR was starting to break out of the South and appeal to a wider audience. NASCAR also saw a changing of the guard. Older drivers retired, and young future superstars were on the rise.

Never was this more evident than at the final race of the 1992 season at Atlanta Motor Speedway. This was the last race for the King, Richard Petty, but that wasn't the only storyline. There were a total of five drivers eligible to win the season title. In a sport where teams of two or more drivers were dominant, driver-owner Alan Kulwicki led one more lap than Bill Elliott (who won that race in Atlanta that day), earning him five bonus points for leading the most laps, and he won the title by 10 points.

Alan Kulwicki also represented the changing face of NASCAR. Kulwicki came from Wisconsin, where he got experience racing on local short tracks. Known as "Special K" and the "Polish Prince" based on his heritage, Kulwicki moved to NASCAR in 1984, when he began his career in the Busch Grand National Series, which had grown into a feeder series for Winston Cup. Kulwicki was part of a group of racers from the Midwest that included such names as Rusty Wallace and Mark Martin. These drivers had raced in the American Speed Association (ASA), a premier short track sanctioning body founded in Indiana that hosted most of its races in the Midwest.

In 1985, Kulwicki moved to Charlotte, North Carolina, and made his first Winston Cup start at Richmond in September for owner Bill Terry. Kulwicki didn't exactly find a warm welcome from his fellow competitors. After all, most of NASCAR's drivers still hailed from the South. Kulwicki was also a college graduate with a degree in mechanical engineering, at a time when most drivers had barely graduated high school. Some, including Dale Earnhardt, had never graduated at all.

In 1986, Terry bowed out of NASCAR, and Kulwicki bought the team and in 1988 won his first Winston Cup race at Phoenix. To celebrate, after the checkered flag, Kulwicki turned his car around and lapped the track backwards, dropping his window net and waving at fans. The celebratory move became known as the "Polish Victory Lap," a practice carried on by many drivers today.

After winning the 1992 Winston Cup championship, Kulwicki quickly gained acceptance from his competitors and fans. His reign, however, was tragically cut short. On April 1, 1993, Kulwicki was killed in a plane crash while returning from an appearance in Knoxville, Tennessee, en route to Bristol, where the Winston Cup Series was racing that following weekend. Three days later, NASCAR raced with heavy hearts at Bristol, with race winner, and Kulwicki's friend, Rusty Wallace honoring his friend with a Polish Victory lap.

That same year also saw the death of another young rising star, Davey Allison, son of legendary driver Bobby Allison.

Young Allison was making a name for himself in the sport in the late 1980s and early 1990s. Allison always drove on the edge, as most successful drivers do. Davey, however, seemed to push it too far on many occasions. Spectacular crashes and emotional finishes marked his career, which was ultimately cut short in a nonracing accident.

Davey Allison made his Winston Cup debut in 1985 at Talladega, finishing 10th. That finish got the attention of many, and in 1986, Allison made five Winston Cup starts. The following year, Allison was tapped to replace Cale Yarborough, who was moving to start his own team at Ranier-Lundy Racing. Allison made history, becoming the first rookie in history to qualify on the front row for the Daytona 500. He would win his first Winston Cup race later that season in a race that changed the sport forever.

The Winston 500 at Talladega Superspeedway was held on May 3, 1987. During qualifying, Bill Elliott won the pole with a speed of 212.809 miles per hour, the fastest time ever officially recorded on a closed track for a stock car and a record that still stands to this day. During the race on lap 22, Bobby Allison spun, exiting turn 4. His car was sent airborne in a spectacular crash that saw the car barrel roll along the frontstretch grandstand fence. Allison was uninjured, but several fans suffered injuries. Davey was ahead of his father and saw the crash unfold in his rearview mirror. The race was stopped, red flagged, for just over two and a half hours. When the race resumed, Elliott was forced to retire with engine failure. During a late race caution, NAS-CAR announced that the race would end 10 laps from the scheduled distance due to darkness. Allison passed leader Dale Earnhardt on the restart and pulled away for his first win.

In the aftermath of that race, NASCAR searched for ways to slow the cars and (they hoped) keep them from going airborne at the two superspeedways, Talladega and Daytona. After trying smaller carburetors, the sanctioning body settled on restrictor plates, plates placed under the carburetor that restrict the air going into the engine, thus lowering the engine power and speeds. Restrictor plates are still in use today at Talladega and Daytona.

The first restrictor plate race in NASCAR's modern era was the season-opening Daytona 500 in 1988. Allison again qualified on the front row but struggled much of the race while father Bobby ran near the front. In the closing laps, the younger Allison was able to make his way toward the front and found himself behind father Bobby. The elder Allison held off his son for his third Daytona 500 win in a race that is among the most legendary in NASCAR. Both father and son would say that the 1-2 finish in that 1988 Daytona 500 would be one of the highlights of their respective careers.

The rest of that 1988 season was forgettable for Davey. Although he ran well at Daytona, he struggled in other races, failing to finish several due to engine failure. His team owner, Harry Rainer, was also struggling and was looking to sell the team. All of that was put on hold on June 19, 1988, when Bobby Allison nearly lost his life at Pocono Raceway.

The Pocono Raceway opened in 1971 and is still one of the few tracks on the NASCAR circuit that remains family-owned. The Winston Cup series held its first event there in 1974, with the sport anxious to race at a track only hours from New York City. Known as the "Tricky Triangle," the unique layout consists of three very different corners and features high speeds and the longest frontstretch in NAS-CAR, at 3,740 feet (Talladega can claim the longest straightaway with a 4,000-foot backtstretch).

During the first lap of the 1988 Miller 500, Bobby Allison crashed nearly head-on into the wall at turn 2. The crash wasn't hard, but seconds after, a car driven by Jocko Maggiacomo running at nearly full speed hit Allison's car on the driver's side full-on. Allison was initially declared dead at a nearby hospital but was resuscitated. He would cling to life in a coma for several weeks. Though he ultimately emerged from his coma and survived, his racing career was over.

The Allison family banner was raised by Davey. The young Allison failed to finish the next three events, but when the Winston Cup Series returned to the site of his father's near-fatal accident at Pocono the following month, Davey finished third. The following week at Talladega, Allison suffered yet another engine failure. In a somewhat dramatic incident later in the race, Davey would take over in relief of Mike Alexander, who was suffering from heat exhaustion and was driving Bobby's car. While the car finished a disappointing 25th, Bobby Allison's car did finish with his son behind the wheel.

Harry Rainer sold his team to Robert Yates in October of 1988. The new team, Robert Yates Racing, struggled with Davey behind the wheel, winning only a few races as the decade ended. It wasn't until a new chief, Larry McReynolds, was hired in 1991 that Allison and Robert Yates Racing became a viable Winston Cup contender. Allison won five races in 1991 and finished third in the final season standings.

In 1992, Allison made a strong statement right out of the box, leading 127 laps in the Daytona 500 and winning the event in dominating fashion. He was a contender in every race early in the season, but a hard crash at Bristol in April left him with a bruised shoulder. In a testament to his toughness, Allison won the next week' s race at North Wilkesboro. He injured his ribs in a hard crash at Martinsville the following week but again raced through the pain to score his third win at Talladega a week later in the same car he had won with at Daytona.

The Winston All-Star Race, the non-points event at Charlotte, was the following week. It was a race that Allison had won the previous year and hoped to repeat. The 1992 All-Star Race was the first held under the lights and featured separate segments. Allison was third in the final segment of the race, behind Dale Earnhardt and Kyle Petty, son of the legendary Richard Petty. On the final lap, Earnhardt and Petty collided, and Allison rocketed past. Petty was able to recover, however, and he and Allison were side by side as they charged toward the finish line. Petty got into Allison as they crossed the line, and Allison's Ford spun into the outside wall on the driver's side. Allison was credited with winning the race but didn't know it until hours later. He was removed unconscious and airlifted to a local Charlotte hospital. Allison suffered a bruised lung and a concussion. The Victory Lane celebration was a somber affair attended by few, as not only was Allison on his way to the hospital, but most of his crew, led by McReynolds, had gone to the hospital, as well.

The crashes and injuries for Allison were far from over. Despite still recovering from his injuries in the All-Star Race, Allison raced in the Coca-Cola 600 the very next week, finishing an incredible fourth. He would win at Michigan in June, and when the series reached its halfway point with the return to Daytona in July, Allison was the series points leader. The following week, the series returned to Pocono. Allison led 115 laps during the first half of the race but fell back into the pack with mechanical issues. While racing back toward the front, on lap 150, Allison and Darrell Waltrip made contact nearing the same turn that almost took the life of Allison's father only a few short years prior. Allison's Ford was sent airborne, with the barrel rolling several times before coming to rest atop a guardrail near the same spot as his father's car did in 1988. Davey was airlifted to the same hospital but survived. He suffered severe head injuries, as well as a broken arm, wrist, and collarbone.

Davey would appear the following week at Talladega wearing dark sunglasses to hide his eyes, which were still horribly red as a result of the Pocono accident. After some prodding from the media in attendance, Allison removed the glasses and smiled, despite the gasps of the crowd. NASCAR rules permit the points in a race to be awarded to the driver who starts a race, so for Talladega, Allison had his cast fitted in a manner that would allow him to shift, and Velcro that allowed him to grip the transmission shifter. Davey started the race but soon turned the car over to a relief driver. The following week, the Winston Cup Series raced on the road course at Watkins Glen, New York, and Allison bowed out, handing the car to a relief driver for the entire race.

Davey made plans to return when the series headed to Michigan in August. Allison had won at Michigan in June prior to his Pocono crash and looked forward to returning. But a tragic event in the week leading up to the race brought more heartache to the Allison family. Davey's younger brother, Clifford, who had hoped to carry on the Allison legacy, was killed while practicing for the Busch Series event at Michigan. Davey raced with a heavy heart to a fifth-place finish and left the track without comment to head to his brother's funeral in Hueytown, Alabama.

Davey would win the penultimate race in the 1992 season at Phoenix, and as the series headed to the final race of the season, Allison needed only to finish fifth or better to win the Winston Cup championship.

That historic 1992 race at Atlanta was famous, but not for Davey Allison winning the title. Allison struggled much of the race, and the story was an epic battle between Bill Elliott and Alan Kulwicki, who swapped the lead for much of the race. Nearing the end, Allison was able to race his way into the top five and was ready to snatch the title, but fate intervened, and Davey was swept up in a crash, ending his hopes for his first title.

Dale Earnhardt and Rusty Wallace dominated the early part of the 1993 season. Allison did win at Richmond but was seventh in points after the traditional halfway point of the season at Daytona in July.

The following week, NASCAR headed to a brand new venue, the New Hampshire International Speedway in Loudon (it would be renamed in 2008 to New Hampshire Motor Speedway). Davey finished third, and the team had renewed hope for a turnaround for the second half of the season.

The next day, Allison accepted an invitation to watch the son of his friend Neil Bonnett test a Busch series car at Talladega. On arriving to watch the test, Davey Allison crashed his new helicopter while attempting to land in the infield. He lingered in a coma overnight before being pronounced dead the following morning.

The year ended with Earnhardt winning his sixth title, as he and driver Rusty Wallace took a Polish victory lap at Atlanta carrying the flags of Davey Allison and Alan Kulwicki.

Whereas the deaths of Davey Allison and Alan Kulwicki both occurred off the track, during the 1990s, there were several NASCAR deaths that occurred on the track. J.D McDuffie was killed during the 1991 road course race at Watkins Glen. A friend of the Allisons and Earnhardt, Neil Bonnett was killed during practice at Daytona in 1994, and Rodney Orr was killed three days later during practice, casting a pall over the 1994 Daytona 500.

The issue of NASCAR safety had always been debated in the sport. NASCAR put many safety innovations in place since its inception. But the argument had always been made that racing was a dangerous sport, and that danger was part of its attraction.

The death of Bonnett affected everyone in the NASCAR garage, but none more perhaps than Dale Earnhardt. Earnhardt and Bonnett were close friends, and Bonnett's death was a blow to Earnhardt.

Earnhardt raced to a seventh-place finish in that 1994 Daytona 500, another loss in a race that had eluded NASCAR's biggest star for many years.

The Daytona 500 and Dale Earnhardt were rivals for all those years. Earnhardt had won races at nearly every other track on the NASCAR circuit and had wins in the Daytona 500 qualifying races and the July race at Daytona, but the 500 eluded him. He led the most laps in the Daytona 500 on several occasions only to come up short of the win due to mechanical issues and even a broken windshield from a strike by a wayward, and unfortunate, seagull.

That's why when Earnhardt finally won NASCAR's biggest race in 1998, it was one of NASCAR's most memorable finishes. As he rolled his famous black number 3 Chevrolet down pit road after the race, crewmembers and fellow drivers lined up to congratulate Earnhardt as millions of fans around the world cheered.

Earnhardt's Daytona 500 win came a year after the 500 victory by a young driver who had become Earnhardt's, and many NASCAR fans', biggest rival.

Jeff Gordon wasn't the first star in NASCAR to come from outside the South. However, Gordon burst on the NASCAR scene at a time when the sport's popularity was building. Born in California, Gordon began racing quarter midgets at the age of six. By the time he was sixteen, Gordon had become the youngest driver to ever earn a USAC license. His family had also relocated to Pittsboro, Indiana, near Indianapolis, in order to provide more opportunities for the young racer.

By the time the 1990s arrived, Gordon had an impressive racing résumé. He was the youngest USAC champion ever, having won the Silver Crown Series title in 1991, and had raced in some lower-level NASCAR events. His main interest, however, remained open wheel racing. Unable to secure a ride in the IndyCar series, Gordon soon had his sights set on stock car racing. By 1991, he was splitting his time in USAC and competing full time in the NASCAR Busch series. He won Rookie of the Year honors in that series in 1991, and three races in 1992, ending the season fourth in points. In late 1992, Rick Hendrick, who was in the midst of building one of NASCAR's most powerful multicar teams, saw Gordon competing in a Busch race at Atlanta. Hendrick signed Gordon a few days later. Gordon made his Winston Cup debut in the famous 1992 season-ending race at Atlanta, finishing 31st after crashing.

Gordon entered the 1993 season racing full-time in the Cup series with Hendrick Motorsports and crew chief Ray Evernham. The

Gordon/Evernham pairing would become one of the most dominant in NASCAR history. Together, the duo won 49 races and three championships in the 1990s.

When he first started out early in the decade, however, Gordon was somewhat of a controversial figure. At the time, most drivers in the upper NASCAR ranks were in their late 20s or early 30s when they began racing in stock car racing's premier league. Gordon began racing in the Cup series full-time when he was just twenty-two years old. As a result, criticism came from drivers who felt that Gordon didn't possess enough stock car experience. His first full year in Cup seemed to suggest just that, as he failed to finish 11 of his 30 races.

Many fans also didn't take too kindly to Gordon. In addition to the fact that he did not hail from the South, Gordon had received his racing experience in the open wheel ranks. Gordon was also handsome, well-spoken, and the more he beat Dale Earnhardt the more Earnhardt's legions of fans disliked him.

With his boyish good looks, charm, and well-spoken manner, Gordon also attracted the attention of marketers. He not only paved the way for a youth invasion in NASCAR, but he set the bar for those new drivers. His polished image and ability to "mind his manners" away from the track was unlike many of the earlier drivers in NASCAR, some of whom not only broke the law moonshine running, but partied away from the track as hard as they raced on the track.

In 1994, Gordon won his first Cup race, the Coca-Cola 600 at Charlotte. It was a win three months later, however, that put many on notice that Gordon was not only in NASCAR to stay, but a force to be reckoned with.

The Indianapolis Motor Speedway and the Indy 500 were still huge in America. There had been talk of a second race at the hallowed speedway, but it never came to fruition. That all changed in 1991.

As NASCAR's popularity was growing, the interest in expansion was in the forefront of the mind of Bill France Jr. In 1991, Indianapolis Motor Speedway president, Tony George, sought approval from the speedway board to add a second race. Once that was secured, he approached NASCAR and a very willing Bill France Jr.

In June of 1992, nine Winston Cup teams tested at Indy. Although there had not been an official announcement, a reported 10,000 fans showed up just to watch. After the test, the speedway began a renovation project that would allow stock cars to race. On April 14, 1993, Tony George and Bill France Jr. unveiled plans for the inaugural NASCAR Winston Cup race, the Brickyard 400, which would take place on August 6, 1994.

Despite some mild opposition from open wheel fans, the first Brickyard 400 was a sellout. It garnered huge attention for not only for the Speedway, but also for NASCAR all around the nation and the world.

Gordon won that first race as thousands of fans cheered their hometown hero. It was his second Cup win, but far from his last. The Brickyard win, however, thrust Gordon into the spotlight and cemented his role as a NASCAR star. Later that same year, Dale Earnhardt won his seventh Cup title, tying Richard Petty; Jimmie Johnson would add his name to the seven-title list in 2016.

The next season, 1995, marked the year the fabled Gordon-Earnhardt rivalry grew. Gordon would win seven races, while Earnhardt won five. Gordon was leading the points as the series headed to the final race of the season at Atlanta. Earnhardt gave everything he could, leading the most laps and winning the race. Although Gordon struggling to a 32nd place finish, Earnhardt fell 34 points behind. Despite his finish, Gordon celebrated his first title as Earnhardt settled for second in the season-ending points tally.

That 1995 season set up the Gordon-Earnhardt rivalry that would last the rest of the decade. Fans chose sides; the older fans with southern roots gravitated toward Earnhardt, while mainly younger fans tended to side with Gordon. Their on-track appearance only fueled that contention—Earnhardt drove a black Chevy, while Gordon drove a brightly rainbow-colored Chevy. Gordon's pit crew at the time was known as the "Rainbow Warriors," which also didn't sit well with Earnhardt and other NASCAR fans.

The truth was that while Earnhardt and Gordon were fierce competitors on the track, off the track they grew to be close friends. Earnhardt referred to Gordon as "The Kid" and "Wonder Boy" and took him under his wing, teaching Gordon the ins and outs of the business of NASCAR.

Gordon and Earnhardt competed against each other 258 times in points-paying Cup races during the rivalry era. Gordon won 52 times, Earnhardt 23. Gordon won three of his four titles in the 1990s, while Earnhardt won four, including his seventh and final one in 1994.

The Gordon-Earnhardt rivalry was far from the only story in NASCAR during the 1990s, but many credit the two drivers with helping to grow the sport from a regional one to the national phenomenon it was by the end of the decade.

Another significant event in the 1990s was the founding of a new top touring series. In May of 1994, the NASCAR Craftsman Truck series was introduced. The series races modified full-bodied pickup trucks. It made its debut in February of 1994 and remains a popular staple on the NASCAR circuit to this day, making it the third of what today is known as the lowest rung on the ladder of NASCAR's top three touring series.

The addition of the Truck series only added to the increase in the popularity of NASCAR during the decade. By the mid-1990s, NASCAR was not only a sport, it was becoming a lifestyle. The attendance at NASCAR tracks grew at a phenomenal rate in the 1990s. Hundreds of thousands of people would converge on a track for a race weekend. The traveling circus that was NASCAR would move from city to city each week, bringing in thousands of support personnel, vendors, teams, and along with the fans, millions of dollars to the local economy. Tickets for many races, especially the high-profile events such as the Daytona 500, Coca-Cola 600, Brickyard 400, and the night race at Bristol, would be snatched up as soon as they went on sale. The races would sell out long in advance, and the seats filled each and every week.

From 1993 to 1998, NASCAR Cup Series at-track attendance alone grew 57% (by 2.2 million) to over 6.3 million, and its top three divisions (Cup, Busch, and Truck) combined grew a staggering 80% (by 4.1 million), to over 9.3 million.[6]

Mainstream media picked up on the growth, and NASCAR stories appeared on the cover of publications such as *Sports Illustrated* and *Forbes*. NASCAR entered the digital age in 1995, launching its website, NASCAR.com, and in 1997 the sport expanded, adding races in markets in Los Angeles, Dallas/Ft. Worth, and a second race in New Hampshire. The following year, 1998, NASCAR added a race in Las Vegas. That same year, NASCAR celebrated its 50-year anniversary led by the naming of its 50 greatest drivers just prior to the season-opening Daytona 500.

Not all of NASCAR's fans were on board with all the changes occurring during the decade. After all, the expansion into new markets not only brought new fans, but attracted major corporations, too. These new sponsors wrote big checks but demanded their drivers become more polished. The days of the grizzled older drivers with grease under their fingernails disappeared, replaced by youthful drivers who not only had the talent behind the wheel, but also had the look and personality that could sell a sponsor's products.

These older NASCAR fans resented the changes. As NASCAR continued its march out of the South into new and larger markets, older NASCAR tracks fell by the wayside. Perhaps no one was more responsible for this than O. Bruton Smith.

Smith built and promoted the races at Charlotte Motor Speedway. A joint venture with Smith and driver Curtis Turner, the track broke ground in 1959. A contract was signed with NASCAR to run the first race on Memorial Day 1960. The race was NASCAR's attempt to stage an event that could directly compete with the Indianapolis 500 during Memorial Day weekend. Construction delays, however, postponed that first race, which was run a few weeks after Memorial Day on June 19, 1960.

NASCAR had raced in Charlotte before, which was (and still is) home to many NASCAR teams since 1954 at the Southern States Fairgrounds track, a half-mile dirt oval. The new Charlotte Motor Speedway was a 1.5-mile paved superspeedway, however, and represented NASCAR's future. The last NASCAR Grand National race at Southern States Fairgrounds was held in November of 1961.

The "World 600" was NASCAR's longest event and remains so to this day. The first 600-miler had 60 entries with 54 cars in the starting field. The race wouldn't run on the same day; it was always scheduled for the Sunday prior to Memorial Day, with Indy always scheduled for Memorial Day proper. This allowed several drivers to compete in both races, and at the time, the Cup points system was forgiving enough that a driver could skip a NASCAR race and still be in contention for a championship; thus, several of the drivers skipped the World 600 to focus on Indy. To date, no full-time NASCAR driver has won the Indy 500 and the World 600 when they are run on the same day, although in 1970 Donnie Allison won the Charlotte race and finished fourth at Indy the following week.

Memorial Day was moved to the final Monday in May starting in 1971, and the Indy 500 found its permanent date on the schedule as the Sunday prior. As a result, no driver attempted the Indy-Charlotte double, as it became known.

In 1992, Charlotte installed lights, becoming the first superspeedway on the NASCAR circuit to do so. The following year, the Coca-Cola 600 (as the race is still known) moved its start time to later in the afternoon with a finish that would come under the lights. In 1994, Indy car driver John Andretti, who had recently announced his switch to NASCAR full-time, became the first driver to attempt a same day Indy-Charlotte double. After finishing 10th at Indy, Andretti was whisked off to Charlotte in time to make the start of the race. He finished 36th, thanks to engine failure, which sidelined him after 220 of the 400 laps.

Others who attempted the same-day double through the years include former Indy car drivers who switched to NASCAR full-time, like Robby Gordon (no relation to Jeff) and Tony Stewart. Kurt Busch was the last driver to attempt the double in 2014. Busch, who had never raced in the Indy car series previously, finished on the lead lap in sixth at Indy but failed to finish at Charlotte due to engine failure.

The Charlotte Motor Speedway hosts two Cup races a year, along with the annual non-points All-Star Race, but the track's fame really grew in the 1970s. Smith had left Charlotte to pursue other business ventures in 1962. With the success of those ventures, Smith began buying shares of stock in the speedway and by 1975 was the majority stockholder. Smith, along with his new general manager of the speedway, H.A. "Humpy" Wheeler, began a bold expansion of the speedway that included more grandstands and luxury suites. In a first for a NASCAR speedway, Charlotte added condominiums overlooking the track that were lived in the entire year.

Wheeler was one of the most brilliant race promoters in NASCAR. He commissioned wild car stunts and full-blown military reenactments prior to races and created a carnival-like atmosphere that drew hundreds of thousands of fans each year.

While Humpy Wheeler was filling the stands at Charlotte Motor Speedway, Bruton Smith wasn't sitting still. Smith began building a racing empire through his company Speedway Motorsports Incorporated. Through SMI, Smith began building an inventory of tracks starting with Atlanta Motor Speedway in 1990. By the end of the decade, SMI owned five NASCAR venues and was the largest track owner in NASCAR. For the first time in the sport's history, ISC wasn't the only corporate player in NASCAR. And Bill France Jr. was none too pleased.

In 1999 ISC merged with Penske Motorsports. Owned by former racer and successful businessman Roger Penske, it had several race tracks in its portfolio. When the decade ended, ISC was back on top, owning more tracks than SMI.

NASCAR was on top of the world at the end of the 1990s. It had never been more popular, made more money, or reached such heights. There were NASCAR branded products everywhere, video games, and comic books. The growth showed no signs of peaking.

The decade ended with the announcement in November of 1999 of a consolidated TV package. No longer would races be broadcast on a hodgepodge of networks from week to week. Instead, the races would be divided among major networks, with one showing a set amount of races in a season before handing coverage off to another.

2000s

The decade began for NASCAR with a historic announcement. In November of 2000, Bill France Jr. announced he was stepping away from day-to-day operations. Few knew at the time that he had been diagnosed with cancer. Mike Helton was named president of NAS-CAR, becoming the first non-France family member to run NASCAR. Helton, who had worked his way through the NASCAR ranks after being hired at Atlanta Motor Speedway in 1980 and spending a year as NASCAR's senior vice president and chief operating officer, remains NASCAR president to this day. Helton is at every NASCAR Cup race, serving as the top official in control of the event.

The future looked bright for NASCAR as the new millennium dawned. Millions of dollars were rolling in thanks to the new TV contract; fans were eagerly snapping up NASCAR branded merchandise and filling the stands each week. New stars were emerging on the track with names like Tony Stewart, Johnny Benson, Ward and Jeff Burton, Terry and Bobby Labonte, Steve Park, Jeremy Mayfield, Ricky Craven, Geoff Bodine, Jerry Nadeau, and Matt Kenseth. Darrell Waltrip, the outspoken driver who thrilled fans with his on-track talent and his ability to raise hackles, raced for the final time in 2000, retiring after 84 career wins and 3 titles.

Several sons of famous racers were among the stars. Richard Petty's son, Kyle, had been racing in NASCAR full-time since 1981 and by the time 2000 came around had won all of his eight career races. While Petty's star was fading, the son of the legendary Dale Earnhardt's was just starting to shine.

Dale Earnhardt Jr. is the younger of Earnhardt's two sons. The elder, half-brother Kerry, raced in various divisions in NASCAR with little success starting in 1998. It was Earnhardt Jr., though, who was just coming into his own in 2000. Earnhardt Jr. had won back-to-back NASCAR Busch series titles in 1998 and 1999. In 2000, he made his full-time debut in the Cup series racing for the team his father had started in 1984, Dale Earnhardt Incorporated, better known simply as DEI. The team ran mainly Busch series races and moved to Cup in 1996. Earnhardt Sr. never raced for his own team in the Cup series, instead remaining with Richard Childress Racing for his entire career.

Earnhardt Jr. won his two Busch titles with DEI and by 2000 was competing full-time in the Cup series for the team, along with drivers Steve Park and Michael Waltrip, younger brother of Darrell. The younger Earnhardt son, born in 1974, began racing at the age of 17 in minor series, first in street stocks, then late models all across North and South Carolina while working as a mechanic at his father's car dealership. He graduated and became part of the Busch series full-time in 1998, making an immediate impact, winning seven races and his first of those two consecutive titles.

When he made his Cup debut in 2000, it looked as though another Earnhardt could write his own legend in NASCAR. Earnhardt Jr. would win two races in his inaugural season, barely losing rookie of the year honors to fellow rising star Matt Kenseth.

As 2001 began, many felt the young Earnhardt's popularity would begin to rise. Nevertheless, it did so for all the wrong reasons, and 2001 would become known as the most horrible season NASCAR ever endured.

The season-opening Daytona 500 can't come soon enough for those who work in, and follow, the sport of NASCAR. The first race of the season is a chance for drivers to start working toward a season title, erasing any troubles they may have had the previous year; new drivers make debuts with new teams, and fans who have longed to hear engines rev during the short winter break finally get their chance.

So it was with the 2001 Daytona 500. The weeks leading up to it were filled with preseason testing, the Winston Open non-points race, and the twin 125-mile qualifying races. Bill Elliott won the pole, and with the debut of the new television package signed in 1999, Fox Sports covered its first Daytona 500. On lap 173, the famous "big one" erupted, involving 18 cars. The term the "big one" refers to the multicar crashes that are typical of the racing at both Talladega and Daytona. The 2001 Daytona 500 "big one" saw young driver Tony Stewart's Chevy become airborne. No drivers were hurt, but the incident stopped the race for an extensive cleanup.

It was the final lap, however, that will be remembered forever.

Michael Waltrip led that final lap, Dale Earnhardt Jr. was second, and Dale Earnhardt Sr. trailed both DEI cars in his Richard Childress Racing Chevy. Entering turn 3 and heading toward turn 4, Earnhardt Sr. appeared to try to block driver Ken Schrader. The ensuing contact sent Earnhardt's number 3 Chevy into the outside retaining wall. It then slid down and came to rest in the grass just below turn 4. Ahead, Waltrip raced to his first win in 462 tries, with Earnhardt Jr. coming home second.

The Earnhardt Sr. crash didn't look particularly bad, but Schrader, a close friend of the Earnhardt family, ran to the car, looked in, and appeared to nearly faint. He began to urgently wave his arms at emergency personnel. Darrell Waltrip, now an analyst with FOX, cheered his brother onto victory from the booth. Almost as an afterthought, he looked toward the Earnhardt crash and said, "How about Dale, is he okay?"

Moments later, with joyful tears running down his face as his brother celebrated his first win, Darrell said, "I just hope Dale's okay. I guess he's all right, isn't he?"

It turned out that Dale Earnhardt Sr. wasn't okay.

Earnhardt was removed from his car after some time. He was loaded into an ambulance and rushed to Halifax Medical Center, less than a mile from the track. TV showed Dale Earnhardt Jr. rushing to be by his father's side, interspersed with shots of Michael Waltrip first celebrating, then looking forlorn and anxious in victory lane.

The ambulance arrived just before 5:00 p.m. Dale Earnhardt Sr. was pronounced dead at 5:16 p.m.

At 7:00 p.m., NASCAR President Mike Helton met with the media and uttered the words that fans and those in NASCAR would remember forever.

"This is undoubtedly one of the toughest announcements I've ever had to make," Helton said. "But after the accident in turn 4 at the end of the Daytona 500, we've lost Dale Earnhardt.

"'NASCAR has lost its greatest driver ever and I've personally lost a great friend,'" he added. "That's Bill France's quote; that pretty much sums it up for the NASCAR community right now."

Earnhardt's death certainly wasn't the first in NASCAR. In fact, the year prior, three drivers had died. Adam Petty, son of Kyle and grandson of Richard, died in a crash at New Hampshire Motor Speedway in May while practicing for the Busch series race. Kenny Irwin Jr. died at the same track two months later during practice for the Cup race, and Tony Roper died after a crash during the Truck Series race at Texas Motor Speedway in October.

Earnhardt's death, however, came on the sport's biggest stage, Daytona, during its biggest race, the Daytona 500. It also occurred at a time when NASCAR's popularity was soaring. His death garnered worldwide media attention for the sport. Many began asking how it could have happened and what NASCAR could or should have done, considering the three deaths the year prior. Fans mourned and gathered outside the speedway in Daytona and the DEI headquarters in North Carolina the days after the crash. The news and the investigation were front-page stories for months.

Behind the scenes, NASCAR knew it had to do something to prevent such a tragedy from occurring again. On the track, NASCAR did what it always did after a driver died—the sport raced on. The tradition of continuing to race after a driver's death had always been a part of NASCAR, so the very next week NASCAR raced at Rockingham, a track in the heart of North Carolina.

NASCAR and law enforcement launched an investigation into the crash. Many press conferences were held in the week leading up to Rockingham, and nearly every detail of Earnhardt's death was revealed. The autopsy showed he had died of a basilar skull fracture, and that his seatbelts might have failed and contributed to the accident. This led Bill Simpson, whose company made the belts used in Earnhardt's car, and many others to resign. Driver Sterling Marlin, whom some blamed for hitting Earnhardt's car and sending him crashing into the wall, received death threats. Michael Waltrip and Dale Earnhardt Jr. were forced to put their grief aside and come to Marlin's defense, publicly absolving him of all responsibility for the accident. Team owner Richard Childress vowed that the famed number 3 would never race again.

A public service was held at a church in Charlotte on February 22, and Dale Earnhardt Sr. was laid to rest shortly after.

NASCAR raced four days later at Rockingham. On lap 3, the fans grew quiet, then stood and raised three fingers to honor Earnhardt. The TV announcers joined in, and the third lap of the race was silent except for the cars racing around the track. This would be repeated at every race for the remainder of the 2001 season.

DEI driver Steve Parks won the race, circling the track in a Polish victory lap with three fingers held aloft outside the window.

The healing had begun.

Two other races in the 2001 season were notable, and many feel that helped the NASCAR community heal. Richard Childress continued to field the car Earnhardt raced. He removed the number 3, replaced it with the number 29, and moved up his young Busch series driver Kevin Harvick to the seat once occupied by Earnhardt.

During the fourth race of the season, only weeks removed from that tragic Daytona 500, Harvick beat Jeff Gordon to the line by inches to win at Atlanta. The crew that originally pitted Earnhardt erupted in a joyous and tearful celebration led by gasman Danny "Chocolate" Myers, a football player-sized man who was as close to Earnhardt as anyone. With tears in his eyes, Myers swept up Harvick and bear-hugged the winning driver.

When NASCAR returned to Daytona in July for the traditional summer race and the halfway point of the season, the death of Earnhardt was still fresh. In the closing laps, Earnhardt Jr. took the lead, followed by Michael Waltrip. The pair finished first and second, respectively, reversing the Daytona 500 finish. Earnhardt Jr. stopped his car on the frontstretch grass near the spot where his famous father had celebrated his first and only Daytona 500 win. Waltrip, along with the crews, soon joined him, and Waltrip and Earnhardt celebrated on top of the winning car.

It was that moment, and that July 2001 win, that many credit with the rise of the popularity of Dale Earnhardt Jr. Many of his father's fans were searching for a driver to cheer for, and they found that in the driver who would become simply Dale Junior, or simply Junior. Junior won two Daytona 500s, in 2004 and 2014, and has been voted the sport's most popular driver since 2003. His net worth is estimated to be over $300 million, and today Junior is the sport's biggest superstar and known inside and outside NASCAR.

While NASCAR continued its investigation into Earnhardt's death behind the scenes in 2001, they also reacted with changes in equipment. Soon after Earnhardt's death, they made mandatory a device that had been only recommended after the deaths in 2000. The Head and Neck Restraint System, better known as the HANS device, is a carbon fiber collar that rides on a driver's shoulders. Lanyards are connected to the

HANS DEVICE

As part of its continuing safety initiative, NASCAR became the world's first major auto racing sanctioning body to mandate the use of an approved head and neck restraint by all drivers on every type of race circuit, beginning with the 2002 season. The HANS Device helps to reduce extreme head motion during accidents and sudden stops. The device is required for use in NASCAR's three national series - NASCAR Sprint Cup Series, NASCAR XFINITY Series, and NASCAR Camping World Truck Series - as well as its regional touring series.

(1) Tethers are attached from the collar of the HANS Device to both sides of the driver's helmet. The purpose of the HANS is to keep the head/neck moving with the torso and not let the head lead or move in opposition to the torso.

1

1

With HANS Device

Upon impact or sudden stop, the two tethers attached to a specially designed shoulder harness help keep the driver's head and neck in a stationary, upright position.

(2) Belts from a six-point safety harness attached to the seat hold the HANS Device in place on the driver's shoulders. The shoulder belts slow down and limit the movement of the torso and the HANS works in concert with the shoulder belts to reduce the movement of the head/neck more than the torso.

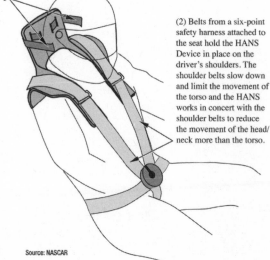

Without HANS Device

Unrestrained, the head and neck of the driver moves forward and/or to the side as the rest of his body and his car decelerate during impact or sudden stop.

Source: NASCAR

SOURCE: *NASCAR*

SAFER Barrier

The SAFER (Steel And Foam Energy Reduction) Barrier technology has been a priority safety initiative between NASCAR, the Indy Racing League, Dr. Dean Sicking and the University of Nebraska-Lincoln since the fall of 2000.

The technology, which helps in reducing the force of initial impact, consists of rectangular steel tubing backed by foam blocks that are installed in front of the race track's traditional cement walls. The process of adding the SAFER Barrier takes approximately two to three months for each track and that time frame consists of the ordering of the specific materials, bending of the steel tubing to fit the radius of the race track's corners, shipment and installation. Once the ordering, bending and shipment are completed, it takes an estimated two to three weeks to actually install the wall.

The SAFER Barrier technology debuted at the Indianapolis Motor Speedway in 2002.

Energy-Absorbing Foam

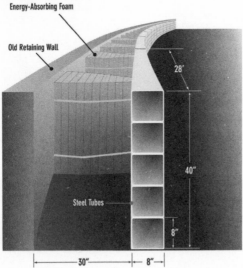

Old Retaining Wall

28'

40"

Steel Tubes

8"

30" 8"

Width

The SAFER Barrier extends 30" from the old wall (may vary from track to track).

Height

40" (may vary from track to track).

Steel Tubes

Constructed of 3/16" thick steel. Each tube is 28' long and 8" x 8" in diameter.

Sections

Each SAFER Barrier section consists of five tubes stacked vertically. Some sections are shorter at the end of the turns.

Closed Cell Foam

2" x 22" x 40" sheets are stacked and bound to create 30" x 40" energy-absorbing pylons (may vary from track to track).

Source: NASCAR

SOURCE: *NASCAR*

back and onto the helmet. Developed in the early 1980s, the device is meant to keep a driver's head from slamming forward during a hard crash, thus preventing the same basilar skull fracture, where the skull is separated from the spine, that killed the three drivers in 2000 and Earnhardt in 2001.

NASCAR only recommended the device at the start of the 2001 season. Earnhardt Sr. refused to wear it, claiming it to be too uncomfortable. After Earnhardt's death, the HANS device was made mandatory, and drivers still wear them today.

Another major safety advance to emerge from the Earnhardt tragedy was the Steel and Foam Energy Reduction, or SAFER barrier. Developed in the late 1990s and early 2000s by engineers at the Midwest Roadside Safety Facility at the University of Nebraska, the barriers are meant to dissipate the energy with a colliding car. Steel tubing is offset by closed-cell polystyrene foam bundles. The SAFER barrier can be installed over existing concrete walls. Prior to the development of the SAFER barrier, much of the attention for dissipating energy had focused on the cars with pieces designed to break away during a crash and in crumple zones, areas of the car that would collapse and absorb energy.

NASCAR had been studying SAFER barriers prior to the Earnhardt crash, but as part of the aftermath, they began to ramp up their efforts. By 2006, every oval track NASCAR raced on had SAFER barriers installed.

Not only was the safety of the drivers addressed, but the safety of crewmembers who serviced the car during a race was, as well. By the dawn of the century, NASCAR pit stops were crucial to winning a race.

During NASCAR's formative years, races were short, and race cars rarely needed fuel or service during a race. As the races lengthened and cars needed service, it was done by the mechanics who worked on the car. But by the 1980s, pit stops were becoming choreographed affairs that quickly serviced a car and helped gain spots on the track. By the 2000s, most pit crews were purpose-built, athletes who trained during the week and whose only job was to service cars during a pit stop on race day.

Today, pit stops are a crucial part of NASCAR racing. *Photo by Greg Engle*

The pits during a NASCAR race are a dangerous place. Several crew-members have been killed, most famously perhaps at Talladega in 1975, when a crew member for Richard Petty, who happened to be his brother-in-law, was killed when a water tank exploded. A crew member died during a race at Atlanta in 1990, when another car spun into the car he was servicing. This accident led NASCAR to institute a pit road speed, anywhere between 35 and 55 miles an hour depending on the track.

After a pit road accident injured three crew members in November of 2001, NASCAR mandated safety helmets for any crew member going over the wall during a race and fire-resistant suits for every crew member working on pit road.

The safety advances didn't slow down as the decade moved along. In 2003, the NASCAR Research and Development Center was opened not far from Charlotte Motor Speedway in Concord, North Carolina. The 61,000 square-foot center replaced a smaller center that opened in 2000. The sanctioning body focused not only on safety issues, but on the future of the sport, as well. The primary focus after the death of Dale Earnhardt became safety on the track as well as the development of safer race cars.

The research and development of that safer race car took five years, and in 2006, the fifth generation of the race car was shown publicly for the first time and came to be known as the Car of Tomorrow, or simply the CoT. The new car featured several safety innovations, including a driver's seat moved four inches more toward the center of the car, larger crumple zones, a roll cage that was moved three inches rearward, a larger windshield designed to produce more drag, and a detached rear wing that replaced a spoiler at the rear designed to help with downforce. The car was two inches taller and three inches wider and had reinforcement that included foam inserts in the doors. There was also an adjustable front "splitter" that replaced the front valance.

NASCAR also debuted a new inspection system that included a laser and a device that looked like a large metal claw that fit over the new cars. There were also common templates, handheld pieces of sheet-metal patterns used to measure a car's body shape and height, with only minor variations between the manufacturers. The new car and the new inspection system sought to eliminate the ambiguity that had existed for years. Prior to the CoT, cars from different manufacturers had different rules and inspection templates.

The new cars were tested throughout the 2006 season and made their competitive debut at the fifth race of the 2007 season at Bristol Motor Speedway. For 2007, however, the CoT was used only at select tracks but then were used for the full 2008 season.

The new cars weren't widely embraced. Drivers and fans complained that they were too boxy, and there was that detached wing. Overall, critics said, the design was too generic, and drivers complained about the terrible handling of the new car.

NASCAR raced on, however, and soon the CoT was seen racing in every Cup race in the series.

Thanks in large part to the safety innovations in the cars and on the track, no driver in NASCAR's top three touring series has been killed on track since that tragic afternoon in February 2001.

The new car and those safety innovations weren't the only changes NASCAR saw in the early part of the new millennium. In fact, many would argue that they paled in comparison to the changes that happened off the track.

One of the bigger changes occurred in 2003, when it was announced in June that longtime series sponsor RJ Reynolds would be leaving the sport. Behind the scenes, the search for a new sponsor had started about the time the new TV deals were being negotiated.

The Tobacco Master Settlement Agreement that involved the four largest tobacco companies, including RJ Reynolds, in 1998 had placed severe restrictions on the activities the tobacco companies could do in terms of marketing. RJ Reynolds began to scale back their involvement in NASCAR, and by 2002, the writing was on the wall. The new millennium was a new era for the sport. The popularity of NASCAR was still growing, and the sport wanted to attract younger fans, a demographic tobacco companies were banned from advertising to. The new TV contracts had an effect, as well. Tobacco companies couldn't buy advertising on TV, which didn't help the networks sell to other potential advertisers.

The new title sponsor announced in 2003 was telecommunications company NEXTEL. NEXTEL would merge with Sprint in 2006. From 2004 until 2008, the top tier was known as the NASCAR NEXTEL Series. From 2008 until 2016, it was known as the NASCAR Sprint Cup series.

It wasn't a completely smooth transition, however. Two other cellular phone companies, Alltel and Cingular, sponsored teams at the time. The title deal nearly fell through as at first NEXTEL wanted the two companies removed and banned from the sport. A compromise was reached, however, that allowed the two companies' sponsorships to continue but banned any new ones from coming into the sport. The ban included Alltel and Cingular, should they decide to exit NASCAR. Verizon would acquire Alltel in 2008 and its sponsorship at Penske Racing (now known as Team Penske). The Verizon brand faced many restrictions due to the ban, and although it tried to market its product on the track at Cup events, the difficulties they faced forced them to withdraw from the sport at the end of the 2010 season.

AT&T took over Cingular in 2007 and in turn wanted to rebrand their Cingular car to AT&T Mobility. NASCAR objected, and soon the battle headed to the courts. NASCAR would eventually prevail, and AT&T was able to stay on the car through the end of the 2008 season, after which it left.

The AT&T lawsuit wasn't the only time NASCAR was in the courts during the first part of the millennium. In 2002, Texas resident Francis Ferko, a race fan who owned stock in SMI, filed a lawsuit against NASCAR and ISC. The suit was filed on antitrust grounds. Ferko was upset that NASCAR wouldn't bring a second Cup race to Texas Motor Speedway, owned by SMI, every season, as he alleged NASCAR had promised when the track opened. The suit was settled out of court in 2004, but the changes it brought about still resonate in NASCAR today.

As part of the settlement, ISC sold Rockingham Speedway to SMI. SMI moved the lone Cup date on Rockingham's calendar to Texas. NASCAR, however, decided a new "realignment" of dates and tracks was needed. They moved the former Rockingham race to Auto Club Speedway near Los Angeles, then gave Texas its second race, but in November. The November date had been Darlington's second race of the season. Darlington was left with only one race in the spring, having lost its historic and prestigious Labor Day race to Auto Club Speedway in 2003. Darlington's spring date fell on Mother's Day weekend, traditionally a poor weekend for racing. Ticket sales plummeted, and many felt the once-legendary track was in its death throes.

ISC didn't abandon the track, though. In 2007 and 2008, the parent corporation poured in $16 million in improvements that included lights. The "Lady in Black," as the track was known, shined. For the next several years, the race sold out, and in 2015, NASCAR moved the race back to its traditional Labor Day weekend. That first "throwback" weekend proved a huge hit. Drivers raced cars featuring the paint schemes of NASCAR's past legends, former TV announcers were used, and the "Lady in Black" shined like never before.

Yet this wasn't the only scheduling change that was taking place during the dozen years between 1999 and 2011, when a period of realignments occurred. Homestead-Miami Speedway had become the last race of the season in 2002. Kansas Speedway and Chicagoland Speedway were added to the schedule in 2001.

Another lawsuit, this one in 2005, would help bring about the 36-race schedule NASCAR runs today. Kentucky Speedway, a privately owned venue near Cincinnati, Ohio, sued NASCAR, claiming the France family had told them they would have one Cup race per season. The lawsuit was dismissed in 2008, then appealed. It was all for naught, however, as SMI purchased the track that same year. In 2001, SMI moved one of its two Cup dates at Atlanta Motor Speedway to Kentucky.

The NASCAR schedule realignment was just a part of the changes in NASCAR during this time. Months after announcing a new title sponsor starting in 2004, Bill France Jr. announced in September of 2003 that he would be stepping down. He introduced his son Brian Z. France as the new chairman of the board and CEO of NASCAR. Bill France Jr. remained active in the sport but lost his battle with cancer on June 4, 2007.

Brian France worked in the family business starting at an early age. His first job was a maintenance worker at Talladega, a job that included janitorial work and mowing grass. He moved up from managing tracks to being part of the team that created the Truck series, to running NASCAR's entertainment division in Los Angeles.

Both Brian and his older sister, Lesa, worked their way up the ranks; Lesa France-Kennedy took over as CEO of ISC in 2009 after working in several roles there.

After his appointment, Brain France wasted little time getting to work. As head of marketing, Brian was instrumental in bringing NEXTEL and Sprint into the sport. One of his first acts as NASCAR chairman was to ban the practice of cars racing back to the yellow. Prior to this, when a caution was called during a race, no matter where the field was, the racing continued until the leader crossed the start-finish line. This resulted in some dangerous incidents and several major crashes. France ended the practice, and when the yellow was displayed, the field was frozen and slowed immediately. This allowed safety workers to get to the scene of an accident quicker.

That change was minor compared to the big change announced at the beginning of 2004.

The Chase for the NEXTEL Cup was revealed during the annual NASCAR preseason media tour in Charlotte in January 2004. The Chase was designed to bring a "playoff" atmosphere to the sport. Some called it the "Matt Kenseth Rule" after Kenseth won the final Winston Cup title in 2003. Kenseth had won only one race in 2003, but his consistent finishes allowed him to be atop the points in the fall, and secure the title before the season ended. Other drivers, such as Ryan Newman, had won up to eight races, yet because of several races where he failed to finish, Newman had no shot at the season-ending championship. NASCAR would deny that the new Chase had anything to do with Kenseth's title in 2003, and Brain France refused to call this new system a "playoff" format.

NASCAR had been looking at ways to make wins more valuable since 2000. With the new Chase format, that became a reality. For the 2004 season, any driver in the top 10 in points and any other driver within 400 points of the top 10 at the end of 26 "regular season" races was in the Chase. While any driver could win one of the final 10 races of the season, only those in the Chase would be eligible for the championship. All the drivers had their points reset, with the top driver starting with 5050 points and subsequent drivers "seeded" at five-point increments behind (5045, 5040, etc.).

Kurt Busch won the first Chase and his first title in 2004, and Tony Stewart won the second Chase in 2005. However, in 2006, a young driver from California began a domination of the sport that saw him win an unprecedented five consecutive titles from 2006-2010.

Jimmie Johnson, an El Cajon, California, native, came into NAS-CAR racing full-time in the Busch series starting in 2000. With a background in motorcycles and off-road racing, Johnson switched to full body late model cars in 1998 and was soon gaining the attention of NASCAR team owners.

Johnson won his first and only Busch series race in 2001 in Chicago. Soon after, team owner Rick Hendrick signed the up-and-coming driver thanks in large part to a suggestion by his star driver Jeff Gordon. Johnson made his full-time Cup debut with Hendrick at the start of the 2002 season. At the time, the Hendrick stable consisted of cars driven by Gordon, Terry Labonte, Jerry Nadeau (who was replaced after a midseason injury by Joe Nemechek), his son Ricky, Ron Hornaday Jr., and David Green.

Johnson won three races in his rookie season and finished second in the championship points the following season. By 2006, he had already won 18 career races. In 2006, he won five races and his first title. For the next four years, Johnson would defend his title, despite the changes NASCAR made to its Chase.

In 2007, the 400 points within the top 10 provision was dropped. The Chase field was expanded to 12 drivers, with all having their points reset to 5000 (higher than drivers outside the top 12). Each driver was then given 10 points for each race win in the season. Johnson won the 2007, 2008, 2009, and 2010 titles.

NASCAR again tweaked the Chase for the 2011 season. For 2011, the top 10 still made the Chase, but positions 11 and 12 were "Wild Cards"—two drivers inside the 11–20 points position with the most race wins. This system remained in place through the 2013 season.

The only exception to the top 12 rule came when the 2013 Chase field had a 13th driver added due to a scandal on the track. The Chase field is set every year after the race at Richmond International Raceway in the fall, and in 2013, the Michael Waltrip Racing team was accused of using team orders to manipulate the outcome of the race and put one of their drivers in the Chase.

Carl Edwards won the Richmond race on September 7, 2013. What happened behind him, however, forced NASCAR to make some hard choices and was talked about for months. With only a few laps to go, Jeff Gordon was racing in a position that would award him a Wild Card spot. On lap 393, however, driver Clint Bowyer, racing for Michael Waltrip Racing, went for a solo spin. A caution was called, and after pit stops, Bowyer's teammate Martin Truex Jr. (the only MWR driver close to making the Chase) had gained enough spots that he was now in the Wild Card spot. Meanwhile, on the same caution driver Joey Logano was able to make up enough positions to take the Wild Card spot from Jeff Gordon.

After the race was over, questions began to surface about the spin from Bowyer. Dale Earnhardt Jr., who was directly behind Bowyer, said the Bowyer spin was "one of the craziest things I've ever seen." Further suspicion came to light when during video replays radio communications between Bowyer and his crew chief could be heard that indicated the two were worried about Truex not making the Chase. Another team communication between MWR driver Brian Vickers and team officials could be heard where a team official wanted Vickers to pit with only two laps to go. This helped to guarantee a Chase spot for Truex.

NASCAR took little time to react, and the next day the sanctioning body handed down its largest fine in history, $300,000, to MWR. They indefinitely suspended the team official who had ordered Vickers to pit and docked Truex and Bowyer 50 points. The penalty knocked Truex out of the Chase, giving the spot to Ryan Newman. The investigation didn't stop there.

Joey Logano had also grabbed a final Chase Wild Card in the closing laps at Richmond. Radio replays seemed to indicate that a driver with a team that had a technical alliance with Logano's Team Penske was told to slow down and allow Logano to pass on the final restart of the race, thus gaining the spot needed to bump out Gordon from the Chase. The evidence, however, wasn't as conclusive as Bowyer's.

NASCAR placed the Penske team and the team the other driver raced for on probation.

The following Saturday at Chicagoland Speedway, all the teams involved were called together for a closed door meeting. After it was over, Brian France announced that in the "interest of fairness" and because of an unfair advantage due to manipulation, Jeff Gordon would be the 13th driver in that year's Chase.

The fallout for the MWR team was devastating. The team, which had its qualifying times disallowed at Daytona in February 2007 after being caught with an illegal fuel additive, was already under a dark cloud. After the scandal at Richmond, a longtime sponsor announced it would be leaving the team at the end of the season. The team struggled for two more years before shuttering at the end of the 2015 season.

In the end, who got in and who was out of that 2013 Chase field didn't really matter. Jimmie Johnson won his sixth title, having captured two of the final ten races, and finishing in the top 5 in five of the others.

The Chase format was further tweaked in 2014 and even more so in 2017. One important change was that the word "Chase" was dropped in favor of simply "playoff." Also, playoff points—which are earned by drivers for winning one of the first two stages of a race (one playoff point) and for winning the overall race (five points)—were added to the mix.

One thing remains the same: There are 16 drivers who make up the final 10-race playoff field. The formula for making the playoffs is pretty straightforward—win the most often and you're in. A race win during the first 26 regular races virtually guarantees an entry into the Chase. The more wins, the higher the driver will be seeded in the playoffs. If there are more than 16 race winners in the regular season, points determine who gets in and who doesn't.

The playoffs for Cup are divided into elimination rounds. Nine of the last ten races are split into three rounds each. In the first round, the Round of 16, the bottom four drivers in points after the third race are eliminated, paring the field to 12. The next round, the Round of 12, pares the field down to eight after the sixth race, and the Round of 8 trims the field to the Championship 4. These four drivers head to the final race of the season, currently at Homestead-Miami Speedway, and their points are all reset. The season champion is the driver who finishes ahead of the others. For the first two years of this new system, the season champions (Kevin Harvick in 2015 and Kyle Busch in 2016) both won their respective final races and the Sprint Cup. Johnson did the same in 2017.

The idea behind this new format is to motivate drivers to win. In the past under the old non-Chase/playoff system, a driver could go with few wins in a season but with consistent finishes and have the title locked up prior to the final race of the season. This is what Matt Kenseth did in 2004, hence the reason some call the Chase/playoffs the "Matt Kenseth Rule."

There has also been some criticism of the Chase/playoffs. However, the truth is that no driver has ever rebounded from outside the top 10 in points to win the season title over the final 10 races of the season. Under the new system, a driver not only has to win, but survive elimination and finish ahead of his rivals in the final race of the season.

One of the other changes that came with the playoffs is the ability for a driver to miss races and still be eligible to compete in the playoffs. Waivers have been granted in the last few years that allow a driver to make the playoff field if they win a race and make it inside the top 30 in points by the end of the regular season. Most notable was the waiver granted to Kyle Busch in 2015, after he suffered horrific injuries in a terrible crash during the season-opening Xfinity race at Daytona International Speedway in February. The crash, into an inside wall that had no SAFER barrier, left Busch with a broken right leg and a broken left foot. He would miss 11 Cup races at the beginning of the season. But in an incredible testament to the human spirit, and physical rehabilitation, Busch returned to the Sprint Cup series at Charlotte in May. NASCAR granted him a waiver, and Busch answered by winning five of the next nine races, three in succession. He made the Chase and won the final race at Homestead to win the Sprint Cup and cap off one of the biggest comeback stories the sports world has ever seen.

Prior to the 2016 season, championships in the Xfinity and Truck series were determined by the points leader after the final race of the season, as it was in the Cup series prior to 2004. However, in 2016, NASCAR introduced a modified playoff format for the Xfinity and Truck series, as well.

The Chase/playoff wasn't the only change NASCAR made to determine its series champion. In 2011, the points structure itself was changed entirely. The Latford System, which awarded points starting at 185 for the winner, and all the way down to 34 points for 43rd place, was abandoned in favor of a new system that simplified the points. From 2011 to 2016, a driver who finished last got one point, while the winner got 40 points, with three bonus points for the race win. Drivers could also get a bonus point for leading a lap and also for leading the most laps.

For 2017, the points system was tweaked. Playoff points were added to the championship points. Now, a driver receives one playoff point for winning one of the first two stages of a race, and five points for winning the race. The top ten finishers in each of the first two stages will be awarded championship points (10 to 1 each), depending on where they finish. At the end of the race, championship points will still be awarded, with the overall race winner getting 40, but the remainder of the field will be scored on a 35 to 2 scale, and positions 36 to 40 receive only 1 point.

Also in 2011, NASCAR required drivers to declare which series they would run for a season championship. They can still compete in any of the top three series but are only awarded points in one (this is why you may sometimes see an asterisk beside a driver's name in the race results with the note "ineligible").

For 2017, the sanctioning body introduced a few new rules. One restricts Cup drivers who want to race in the Xfinity and Truck series. Drivers in the top-tier Cup series with more than five years of experience can only enter ten Xfinity races and seven Truck series races in a season. In addition, they will not be allowed to race in any of the final eight races in those series that comprise the championship Chase for the respective series.

Only weeks prior to start of the 2017 season, during their annual media briefing at the R&D Center, NASCAR announced that not only will they now have full-time medical crews traveling with the Cup series all season, but that the days of riding around earning points were over once you suffer damage caused by contact on the track.

Prior to 2017, if a car was damaged in an on-track incident, it could go to the garage, repairs made, and that car could reenter the race. They were of course many laps down and had no chance of winning, but as long as they could maintain the minimum speed dictated by NASCAR for that particular track, they could ride around in hopes of gaining a position or two.

Starting in 2017 (for all three of the top touring series), if a car is forced to the garage with damage caused by contact on the track, they cannot return. But if they pit, the crew has five minutes to effect repairs. Those repairs, though, cannot include replacing any sheet metal. If the car can continue at minimum speed, fine; if not, at the end of the five minutes, they must either go back out and repit and get another five minutes or withdraw from the race.

None of this applies to mechanical or electrical issues, however. If a car suffers a breakdown without on-track contact, the crew can repair it in the garage with no penalty, and the car can return to the race.

Other changes that NASCAR has introduced since the dawn of the new millennium include physical changes to the cars themselves.

Since its introduction, NASCAR's CoT wasn't exactly embraced by all. Its boxy shape, the odd detached rear wing, and the front splitter with its wire braces meant the car bore little resemblance to its stock counterpart. When Kyle Busch won the first race with the CoT car at Bristol in March of 2007, he wasn't all that pleased with the new car. "I'm still not a fan of these things," he said in victory lane. "They suck."

There were tweaks to the design. In 2011, the awkward-looking front splitter was modified to give the nose of the car a cleaner look, and the wing was replaced by a more traditional rear spoiler. The 2012 season also marked the first season NASCAR abandoned old-style carbureted engines in favor of fuel injection in the Cup series—a move that brought even more "stock" back into the sport. The introduction of fuel injection came two years after NASCAR announced a move to the more environmentally friendly Ethanol blend fuel, E15, in its three national touring series as part of a move to become more "green." The use of an overflow vent meant to catch fuel that didn't make it into the car was also new.

Even more "stock" was brought back in 2013.

While the CoT racers were the safest in NASCAR history, they looked nothing like their counterparts on the street beyond having manufacturer badges. That all changed after 2012, when the Generation 6 Cup car was introduced. Carbon fiber hoods and decklids shaved weight off the cars, and bigger roof flaps were among the changes, as well. The roof flaps deploy when a car spins backwards and are used to help keep the car on the ground. One of the many criticisms of the CoT was its tendency to get airborne in a crash.

The biggest change with the Gen 6 car, as it came to be called, was the ability to have parts of the body, mainly the nose, more closely resemble their street counterparts. The new car design was a hit with competitors and teams and made its debut in the 2013 Daytona 500.

The Xfinity Series had its own CoT. It was introduced in 2010, and it too got upgraded. In 2015, the new Xfinity car was introduced. These cars are different than those used in the Cup series, whose models are based on sedans. The latest Xfinity cars represent American sports cars like the Chevrolet Camaro and Ford Mustang and enjoy the same kind of universal appeal the Gen 6 Cup cars do.

NASCAR was still riding a wave of popularity in the middle of the first decade of the century. As the healing process over the loss of Dale Earnhardt was well underway, plans emerged for NASCAR to create a Hall of Fame, something that all the major professional sports in America had. In 2005, NASCAR began taking bids from cities that would be interested in hosting the NASCAR Hall of Fame. The cities that prepared bids included Atlanta, Daytona Beach, and Charlotte (Richmond and Kansas City were also on the list but not among the finalists). Charlotte was the clear favorite, as the majority of NASCAR teams had their headquarters in the Charlotte area (and still do).

Charlotte was officially announced as the site of the NASCAR Hall of Fame on March 6, 2006, and ground was broken in January the following year. The NASCAR Hall of Fame in downtown Charlotte opened to the public on May 11, 2010. The first class was inducted several weeks later. Every year since its opening, five inductees are chosen from a field of nominees, by a committee made up of industry representatives, including track owners and drivers and members of the media. A fan vote is also considered.

Around the time of the Hall of Fame announcement, events outside the sport also began to affect NASCAR.

The NASCAR Hall of Fame in downtown Charlotte opened to the public on May 11, 2010.
nascarhall.com

When the American economy began to falter in late 2007 and with it people's level of disposable income began to dwindle, fans of all sports, including NASCAR, found it difficult to afford tickets to see an event in person. This, combined with the fact that fans could watch their favorite sport on the Internet, meant that not only did TV ratings drop, but there were fewer and fewer "butts in the seats," as well. NASCAR tracks that could once fill over 100,000 seats began to see a decline. Richmond International Raceway had sold out every consecutive race since 1992. That ended in 2008, and that track has not had a sellout since.

The recession near the end of the first decade of the new millennium meant that people were more concerned about simply paying their mortgage, not buying tickets to a professional sporting event. This, combined with the traditional practice of some communities surrounding major NASCAR tracks, led to empty grandstands. There was a time, up until the recession, that hotels near NASCAR tracks inflated rates when a race was in town. They also required minimum stays. Prior to the recession, NASCAR fans would pay the inflated rate, and stay the minimum number of nights, to see a race. That ended when the recession hit.

As a result, NASCAR tracks struggled during this time. Richmond International Raceway, which had sold out every consecutive race since 1992, has not had a sellout since 2008. A ticket to the Daytona 500, once coveted and sold out at least a year in advance, became widely available, as did tickets to other races. The tracks responded by first lowering ticket prices, and then lowering seating capacity.

Starting in 2013, Richmond removed grandstands in the third turn and the backstretch to reduce seating capacity from 112,000 to 60,000. That same year, Talladega cut its seating from 108,000 to 78,000, and Michigan from 84,000 to 71,000. Atlanta removed 17,000 seats; Dover 17,500; and Charlotte 41,000 seats in the winter of 2014, bringing its seating capacity to 89,000.

ISC had a total of 1.1 million grandstand seats in 2007. By 2016, that number had dropped an estimated 30 percent to around 760,000.[7]

Perhaps the most celebrated of these grandstand reductions came at one of the most iconic racetracks in NASCAR, Daytona International Speedway. The Daytona Rising Project, announced in 2012, had its groundbreaking just prior to the July race in 2013. The $400 million renovation completely removed the backstretch grandstands, losing 46,000 seats. The seating capacity dropped to 101,500.[8]

The new Daytona International Speedway, called a "motorsports stadium," officially opened in 2016. It includes "injectors" (large

The $400 million Daytona Rising renovation completely removed the backstretch grandstands, losing 46,000 seats. The seating capacity dropped to 101,500.
Photo by Greg Engle

entryways) and four sponsored concourses and a fifth currently unsponsored, containing "social areas" with bars and restaurants. Seat width was increased, and 60 luxury suites now tower over the frontstretch. The 2016 Daytona 500 marked the first time many NASCAR fans got a glimpse of the new Daytona, and for the first time in many years, the 2016 Daytona 500 was officially sold out.

The new Daytona International Speedway debuted in February 2016. *Photo by Greg Engle*

During this time, TV ratings also began to dwindle around the time the recession hit. Once an important measure of the sport's popularity, NASCAR's TV ratings began double-digit declines starting in 2007. This decline mirrored that of most other televised professional sports, as well as general TV network viewing, thanks in part to the advent of high-speed Internet. There was also the argument that attention spans were getting shorter and investing three or more hours to watch an entire game, or NASCAR race, just wasn't happening.

NASCAR responded by offering products on its website that allowed fans to either complement the live TV broadcast or to see and keep up with an entire race completely online. In 2013, NASCAR opened its Fan and Media Engagement Center at its headquarters in Charlotte. The first of its kind, this center now uses analytics from traditional, digital, and social media to provide real-time response to fans and the industry. Still, television networks haven't shied away from NASCAR. In July 2013, NASCAR announced the signing of a 10-year rights agreement with NBC Sports Group, granting NBCUniversal exclusive rights to select national touring series races and other live content starting in 2015 and lasting until 2024.

Just over a month after the NBC announcement, NASCAR and FOX Sports announced an extension and expansion of their multi-platform rights agreement from October 2012 through 2024.

Today, Fox Sports handles the Cup races in the first half of the season, up until the July race at Daytona, while NBC broadcasts the second half of the season.

Though TV ratings may be down, many will argue that this decline doesn't mean people aren't watching races. Brian France said at the end of the 2016 season that he is very pleased about NASCAR's position in sports.

"The audience isn't going away at all," France said. "It's sliding to different places, consuming in different ways."

"I would tell you some other leagues that have 30 percent drop-offs, they didn't lose 30 percent of their audience from one moment to the next; that audience is just sliding and consuming in some different ways," he added. "Our digital consumption is off the charts.

"I watched the other day, as an example . . . a Duke Blue Devils game, after they had played, in like a six- or seven-minute recap of the game," France said. "It was pretty good. Maybe it was a little longer than that. I didn't watch the game on TV, but I watched it fairly intensely with my laptop.

"Sports, in the end, us included, will always have a huge, big audience," he explained. "So whether ratings are sliding over here, spiking at times over here, that will all work out."

In recent years, NASCAR has made changes that have outcomes that may not be seen for some time and some that have changed the way NASCAR races, bringing more excitement than ever.

One of the first changes that impacted fans and competitors was the revamping of the qualifying procedures starting in 2014. Prior to that year, qualifying consisted of a single-car session where one car would make a lap, or two. The lineup was then set fastest to slowest. The Cup field of 43 cars was set this way with provisional starting spots allotted for past champions if there were more than 43 entries. This type of qualifying was the norm for most of NASCAR's history. While it did succeed in setting the lineup, the excitement for the fans was minimal.

NASCAR implemented an open group qualifying procedure at the two-road course races in 2013, but the changes for 2014 were real game-changers. A new knockout-style qualifying procedure was debuted at every track with the exception of the Daytona 500 and the All-Star race.

In 2012, NASCAR began dissecting the rule book line by line. All areas of the sanctioning body were examined in order to modernize the way the sport is governed. The goal is to simplify the rules and to allow for more objectivity. Part of this process included moving the responsibility for rules formation from the competition side to research and development. The penalty and appeals process was redefined and pit road officiating was streamlined.

Prior to the changes, a NASCAR official could be seen by every car in the pits during a race. Under the new system, there are fewer pit road officials. Instead, an extensive system of cameras controlled by NASCAR officials in a trailer are used. The inspection system was also made more efficient and consistent across all three top touring series.

These are just some of the changes to the competition. NASCAR promises there will be more in the years ahead.

Perhaps the biggest game-changer, though, especially where teams are concerned, came in January 2016 when Brian France announced the implementation of the Charter system.

NASCAR has long resisted unionizing their teams and drivers, careful to call them independent contractors. The Charter system, while not a union, seems about as close as they will come.

The Charter system seeks to give teams a better certainty about the business of NASCAR and an increased opportunity to work closely with NASCAR. The agreement is for nine years, and there are 36 charter teams spread out among the organizations. The initial number of Charters given wasn't based on a hard number. NASCAR looked at teams that have a long-term commitment to the sport; they saw which teams had attempted to qualify for every race in the prior three years.

A Charter guarantees a team entry into a race, although qualifying still determines where they will start. In addition, the NASCAR Cup field was pared down to 40 cars from 43. This means that 36 entries will be for Charter teams, while the remaining four will be for "open" teams without Charters.

Several of the larger organizations had teams that didn't qualify for a Charter under the three-year benchmark. These teams were allowed to purchase Charters from other teams. Joe Gibbs Racing, for example, purchased a Charter from the defunct Michael Waltrip Racing for its fourth team, as did Stewart-Haas Racing, which bought the other MWR Charter for its fourth team.

Charter teams are held to a minimum performance standard. If a Charter team finishes in the bottom three of the owner standings among all 36 Charter teams for three consecutive years, NASCAR has a right to remove the Charter. No one outside NASCAR and team executives knows how much a Charter costs; it's generally accepted that it is well up in the seven-figure range. However, team executives are quick to say that since Charters give them a stronger voice with NASCAR and a guaranteed starting spot for every Cup race, the Charter investment will pay for itself.

"Charters are going to be a long process for us," Brian France said. "We got it done. That's the most important thing.

"But forget the value part of it," he added. "The things that we're going to be able to do are going to take many, many years to achieve our end goal, which is to really lower costs in the industry by working with our interests better aligned with the teams, that's number one, which will affect their values going forward.

"Some of the things are not going to change, of course," France said. "Like you got to compete at a high level, you've got to get sponsorship, you've got to have a manufacturing relationship. Those things don't change, nor should they.

"But the benefit of collectively working together in particular to get the rules packages . . . [it equals], more exciting racing for the fans, and much lower cost over time, it's the hardest thing in racing to do," he added. "And frankly, few ever achieve it, or at least consistently achieve it.

"We are after that deal. We couldn't do that without the teams aligned together with us like they are."

In 2015, NASCAR received another avenue to better communicate with its competitors. While not a union, Cup drivers formed the NAS-CAR Drivers Council, a committee of the sport's drivers that regularly meets to discuss competition issues that are presented to NASCAR. NASCAR welcomed this communication, and Brian France attended his first meeting in 2016.

"For the last probably 10 years we didn't even know what a good show was," Dale Earnhardt Jr. said in a press conference shortly after the first Drivers council meeting in May of 2015. "The drivers had an opinion what a good show was, NASCAR had an opinion what a good show was, and it might not still be the same thing."

While there continues to be a debate about how many people were watching the racing (and how), on the track many agree that the racing has never been better. With new rules in place that make the cars more competitive, the finishes in NASCAR have been historically closer (the finish of the 2016 Daytona 500 was the closest in history).

One factor that some say has hurt NASCAR in the past few years has been the lack of a real superstar. Dale Earnhardt Jr. continues to be the sport's most popular driver, but he has not won a title or come relatively close; and he is now over forty, an age when some athletes begin looking at retirement. Jeff Gordon retired from driving in 2015, although he remains as part team owner and TV analyst (and famously got back into the car as a substitute for Earnhardt Jr. for several races in 2016), and Tony Stewart retired after the 2016 season.

However, there are young stars who are rising. Some come from NASCAR initiatives such as NASCAR Next, a program that recognizes young talent, and the Drive for Diversity (D4D), which recognizes talent from diverse backgrounds. Mexican-born driver Daniel Suárez came from NASCAR Next and won his first Xfinity series race in 2016. He went on to win the NASCAR Xfinity Series title that year, becoming the first foreign-born driver to win a title in the top touring series. Darrell "Bubba" Wallace Jr. came from the D4D program and on October 26, 2013, won a Truck series race at Martinsville, becoming the first African-American driver to win in one of NASCAR's top touring series since Wendell Scott in 1963. Aric

Almirola, of Cuban descent, also participated in the D4D program and won the Cup race at Daytona in July of 2014.

Other rising stars come from families with a history in NASCAR. Chase Elliott, son of the legendary Bill Elliott, won the NASCAR Xfinity title in his first full year in that series. In 2016, he moved to the Cup series, driving the famous Hendrick Motorsports 24 for the retiring Jeff Gordon.

Chase Elliott, the son of the legendary Bill Elliott, is quickly becoming a fan favorite.
Photo by Greg Engle

Ryan Blaney's father Dave Blaney raced in NASCAR's top series for over a decade, entering 473 races and retiring from the Cup series in 2014. Young Blaney scored four wins in the Truck series in 58 starts and four wins in 46 starts in the Xfinity series. He moved up to the Cup series in 2016, racing full-time behind the wheel of the legendary number 21 Wood Brothers car.

These drivers, along with other young drivers such as Chris Buescher, Erik Jones, Austin and Ty Dillon (grandsons of team owner Richard Childress), and Kyle Larson, are all drivers whose stars are still rising.

While NASCAR may not enjoy the popularity it once had, it's far from over. The sport is changing and modernizing, but in many ways the same elements that Bill France Sr. promoted all those years ago when he stood on the sandy beaches in Daytona are still there. Most Sundays, you can find a group of talented drivers strapping into cars that still resemble those driven on the streets ready to see who will survive and hoist the trophy when the checkered flag falls. The legacy and history of NASCAR continues.

Brian France was quick to say why he feels it's important for everyone to know the history, and what the legacy carried forth from his grandfather and father is: "It's about the safest, most competitive form of racing in the world," he said. "It's that simple.

"It evolves," he added. "The drivers, the equipment, the manufacturers, the speedways, evidenced at Daytona this year, all evolve. That's normal in sports. But that's the premise of it. The premise of it also is it has to be a fair, balanced playing field where teams who compete and follow the rules have a real shot to win.

"Good, smart parity between the manufacturers and others," France said. "We have that, too. Evidence of that [is] the three manufacturers all in the Chase final four [in 2016]. That's really about as simple as it gets."

CHAPTER 2
LET'S GO RACIN'

SO JUST WHAT is it that you are watching? Whether sitting on your comfy couch at home or in-person at a track, you could be watching a race in a number of the series NASCAR manages. Most often, you'll be watching the Cup series, the premier top series that is the most recognized. However, races in the Xfinity series and the Truck series are broadcast as well and are often companion events to Cup races at the same track during the days leading up to the big Cup race.

You might even visit a local track and see NASCAR officials and NASCAR banners and watch a NASCAR-sanctioned race. The truth is that NASCAR is everywhere today, not only on national TV, but also sometimes surprisingly close to home.

So which series does NASCAR oversee, and where do they race? What are the big races, whom do you cheer for, and just what goes on during a NASCAR race?

Today, NASCAR runs three national series, four regional series, one local grassroots series, and three international series.[1]

The three top touring series are the best known, and most watched. The top of this tier is the Cup series. It's been known as the Grand National Series, the Winston Grand National Series, the Winston Cup Series, the NEXTEL Cup Series, and the Sprint Cup Series since its inception in 1947. For 2017, the series underwent a new name change, as Sprint departed and Monster Energy took over the Cup series entitlement rights and the All-Star race in May. The series is currently known as the Monster Energy NASCAR Cup Series. No matter what, the Cup series remains the top of the tier, the series most drivers aspire to reach, and the series most non-NASCAR fans first see.

The Cup series currently contests 36 points-paying races a season at 23 tracks in 20 states. The official season lasts from February to November. There are also two non-points races. "Points-paying" races refer to those where drivers earn points toward the season-ending championship. The two races that don't earn a driver any championship points are the Clash at Daytona (known until 2017 as the Sprint Unlimited), an invitation-only race one week prior to the Daytona 500, and the annual All-Star Race, which includes a qualifying race. Up until 2017, the Duel Qualifying races at Daytona that set the field for the Daytona 500 during the week leading up to the race were non-points events. Starting in 2017, points will be awarded for the Duels.

The 36 points-paying races currently begin at Daytona in February and, as of 2017, end at Homestead-Miami Speedway in South Florida in November. There are two road courses on the Cup schedule, one in California and the other in New York. There are also two superspeedways—Talladega (2.66 miles) in Alabama and Daytona (2.5 miles) in Florida. Eight tracks are 1.5 miles in length, two are two miles in length, two are one mile in length, and five are of varying lengths: Bristol Motor Speedway in Tennessee is a .533-mile oval. Darlington Raceway in South Carolina is 1.366 miles, Martinsville Speedway in Virginia is .526 of a mile, New Hampshire Motor Speedway in Loudon is 1.058 miles, and Richmond International Raceway in Virginia is .75 miles in length.

The second-tier series is currently known as the Xfinity Series. It was formed in 1982 when NASCAR brought together the regional racing series known under the banner of the Late Model Sportsman Division into a new series. Rising costs made it harder and harder for regional racers to compete on a weekly basis at bigger races with bigger purses. This new series was first known as the NASCAR Budweiser Late Model Sportsman Series.[2] Many of the races were held at historic venues like

Hickory in North Carolina and South Boston, Virginia. Other events, however, were held as companion events to Winston Cup races at Daytona, Richmond, and Bristol, and that trend continues to this day. This series was rebranded in 1984 to the Busch Grand National Series and the NASCAR Busch Series Grand National Division the following year. It would become the NASCAR Busch Series in 2004 and remained that until 2008, when Nationwide Insurance took over as the title sponsor and the series became the NASCAR Nationwide series. In 2014, Nationwide left the series, and Xfinity, a subsidiary of Comcast Cable, became the title sponsor.

Today, the NASCAR Xfinity Series is considered a "feeder" or development series. Almost all of the races are companion races to the Cup series with shorter race lengths and using race cars with less power than those raced in the Cup series. There are stand-alone events in the Xfinity Series on tracks the Cup series doesn't race on, like Iowa Speedway and the road course at Mid-Ohio and Road America. For 2017, there will be a total of 33 Xfinity Series races with five stand-alone events at 25 tracks.

Although many of the Cup stars race in the division each week, many of the drivers are under the age of 25 and are gaining experience in the Xfinity series before moving up to Cup. Despite the presence of Cup drivers, the yearly championship is limited to those who declare prior to the season that they will only compete for points in the Xfinity Series. Thus, Cup drivers can race Xfinity races but aren't eligible for points, and that's the same for Xfinity drivers and Truck Series drivers.

Starting in 2016, the Xfinity Series title will be decided via a playoff-style format. The top 12 drivers will be seeded based on wins during the first 26 races, with the highest among those who have not won a race making it in on points. Like the Cup playoff, there are three rounds. First is the Round of 12, pared down to eight after three races, then the Round of 8 will become four after the next three races, with the Championship four contesting the title at Homestead, and the highest finisher winning the title.

The third of the top touring series is the Truck Series. The series races modified full-bodied pickup trucks. It was introduced as the Super Truck Series in May of 1994. The first race was held in February of the following year as the rebranded Craftsman Truck Series. Many of NASCAR's top teams put resources behind the new series, including Richard Childress, Roush Racing, and Hendrick Motorsports. In its first years, Truck races were held mainly in the Western United States but eventually expanded and now races across the country and in Canada.

Today, young drivers wanting to move up the ranks to the Cup series usually start their climb in the Camping World Truck series, as it's known today. Several of the Truck teams, including Brad Keselowski Motorsports and Kyle Busch Motorsports, are now owned by Cup drivers. Unlike in the Xfinity Series, however, there are many drivers who race only in the Truck series and are quite happy to remain there. Drivers like Matt Crafton, Timothy Peters, and Johnny Sauter run only in the Truck Series. There are still plenty of youngsters working their way up the ladder, but today the Truck Series has a larger complement of regulars than the Xfinity Series, and in the opinion of many fans, the Trucks stage some of the best racing in America.

"I'm very happy to finish my career out here," veteran Truck racer Matt Crafton told me. "I mean, yes, I would like to race the races in the Xfinity Series or run Cup races if I was on the right equipment. But if you're not going to be in the right equipment, you're not going to be able to contend to win. I really don't care to do it. I ran some Xfinity races whenever I drove for RCR and had Menard's on the car there and finished third a couple times and ran very good, but the Camping World Truck Series is such great racing, and I don't just say it because I'm racing in it, but there's so many fans out there, true race fans. It's the best racing in NASCAR.

"I love to be a part of that," he added. "And it's short enough races where we put on a great show from start to finish, and you can still make a living, get to drive a race car and have 23, 24 weekends to do it."

A new sponsor took over the title sponsorship in 2009, and today the NASCAR Camping World Truck series races at 23 events, many as support races to Cup events, but many as stand-alone events including, starting in 2013, a dirt track event at Eldora Speedway owned by three-time Cup champion Tony Stewart.

The Truck Series has also tried some eye-opening gimmicks through the years. In the beginning, the Truck races were too short, and the trucks didn't need fuel, so there were no traditional pit stops during a race. Instead, NASCAR instituted a five-minute break known as "half-time," a stoppage period where teams could make adjustments to their trucks. The Trucks weren't allowed to change tires during a race except if a tire was damaged or flat.

Competition cautions, a caution period called by NASCAR on certain laps and usually made known to teams prior to a race, were introduced in 1998, ending the halftime break. The following year, full pit stops were allowed, but only two tires could be changed at a time.

Perhaps the most popular rule in the Truck Series was the use of the overtime rule or, as it's commonly known, a "green-white-checkered flag" finish. In NASCAR's other top touring series prior to 2004, if a caution came out near the end of the race and lasted for a period beyond the scheduled end of the race, the race ended under caution, with the winner being whoever was leading at the start of the caution.

However, in the Truck Series, the race had to end under green. If a caution happened in the final three laps of a race, the field was held under caution until the track was clear, then set loose for three laps, a green flag, followed by a white flag, and finally the checkered flag. If something happened on the first green flag lap to cause a caution, the field was reset and the green-white-checkered procedure was repeated, as many times as it took. One Truck Series race in 2004 had an additional 14 laps added until a winner was declared.

The green-white-checkered flag finish was so popular that NASCAR adopted it across all three of its touring series in 2004. The biggest change when it was adopted across all the series was that there would be only one attempt at an overtime finish. This was changed in 2010 to allow for three attempts, and in 2016 an "overtime" line was added.

The "overtime" line, different at each track, usually marks the halfway point of a track and the point where officials determine there has been a "clean" restart. If the leader passes this line and there are no issues, the race will end with the next flag, whether that flag is the yellow for a caution where the field is frozen and slowly rolls to the finish or the checkered flag for the winner and allows racing to the finish. However, if a caution is displayed prior to the leader reaching the overtime line, the field is reset, and up to two more attempts are made at a green-white-checkered flag finish.

There were two other notable changes for the Truck Series in 2016. The first involved the introduction of a "caution clock." This 20-minute timer counts down, and if no caution period happens naturally, the caution will be displayed. The clock isn't used during the final 20 laps (10 at Pocono and road courses) and during the dirt track race at Eldora. The "caution clock" was dropped as NASCAR instituted the three-stage format, which will be discussed in detail later.

The other change for the Camping World Truck Series was the introduction of its own modified playoff system. The entry requirements are the same—wins combined with points determine the seeding of eight drivers who comprise the Truck Chase grid. Two are eliminated after each of the two rounds (of the three races) leading up to the finale at Homestead, leaving four drivers to compete for the title; and, like the other playoff formats, the championship winner is the highest finishing driver among the four.

The top three series get the most attention, but NASCAR has a hand in many racing series in North America and Europe. Therefore, while a fan might not be near a major NASCAR venue, it's a pretty good bet that a NASCAR-sanctioned race is happening not far from their hometown on any given weekend.

Under the Home Tracks banner, NASCAR sanctions and oversees several stock car racing series:

The K&N Pro Series East began as the Busch East Series and Busch North Series in 1987. The series races primarily in the eastern part of the United States.

The K&N Pro Series West was started in 1954 as the Grand National West then the Winston Grand National West Series with the intention of allowing drivers who couldn't afford to race in the East a chance to race at the Grand National level. Today, the series races primarily in the western part of the United States.

Both K&N Pro Series share similar rules and race cars and race at short tracks and road courses all over America. Other series include the Whelen Modified Tour, which races open-wheeled modified stock cars in the Northeast; and the Whelen Southern Modified Tour, which mirrors the Modified tour with races in the South. Both tours merged into one for 2017, and the unified Whelen Modified Tour had plans for 17 events in 2017. There is also the Whelen All-American Series, which allows NASCAR to sanction tracks, races, and points championships at a network of tracks, both dirt and pavement, across America and Canada.

NASCAR has also gone outside the US. They sanction a stock car series in Canada once known as the Canadian Tire Series. It gained a new sponsor for 2016 and is now known as the NASCAR Pinty's Series. The Toyota Series races in Mexico, while the Whelen Euro Series, which entered into an agreement with NASCAR in 2012, races stock cars in Europe.

NASCAR even sanctions an online race series. The NASCAR iRacing.com Series debuted in 2010, and it isn't unusual to find drivers such at Dale Earnhardt Jr. occasionally racing against others in an online oval race in cyberspace.

As of 2015, NASCAR sanctions more than 1,500 events in 11 series at 110 tracks in 37 states, Mexico, Canada, and Europe.[3]

Demographics

So who are NASCAR fans? The demographics for each of the top three touring series vary somewhat. However, based on stats from a survey conducted by Nielsen Scarborough in 2015 and provided by NASCAR, for the Cup series, the demographics are as follows:

- 62 percent of fans are male, 38 percent female
- 2 out of 5 are in the 18–44-year-old range
- 23 percent are multicultural, and 3 out of 5 are employed full- or part-time
- The average household income is $71,000
- 2 out of 3 are homeowners
- 1 out of 3 has children
- 1 out of 2 has some college or beyond.

Most NASCAR fans still live in the South (41 percent), followed by the Midwest, (24 percent), the West (20 percent), and the Northeast (15 percent).

The top five NASCAR markets (by the number of people interested in the sport):

1. Los Angeles
2. New York
3. Chicago
4. Dallas
5. Atlanta

Where the NASCAR Cup Series races

The top-tier Cup series stages 36 points-paying races at 23 tracks in 20 states. The tracks range in length from .0526 of a mile (Martinsville, VA.) to 2.66 miles (Talladega, AL). How are there only 23 tracks for 36 races?

That's because NASCAR races at some venues twice in the same season. They begin the season at Daytona in February (and mark the halfway point of the season there in early July) and as of 2017 finish the season in November at Homestead-Miami Speedway in South Florida.

Many of the races in the spring part of the season are at tracks the series will visit again in the fall. Phoenix, Martinsville, Texas, Richmond, Talladega, Dover, and Charlotte are among spring tracks that have two races in the season—one in the spring and the second in the fall.

The part of the season known to many as the "summer stretch" includes two races at Michigan International Speedway (June and August) and Pocono Raceway (June and July). There's also a second race at Bristol (the first is in April, the second in August) and races at Kentucky Speedway, Indianapolis Motor Speedway, and both road course races (Sonoma in June and Watkins Glen in August). The first event at New Hampshire Motor Speedway is held in July (the second is in September; however, starting in 2018, the September race at New Hampshire will go to Las Vegas Motor Speedway, which will have two Cup races per season starting in 2018), and the summer stretch ends with the Labor Day weekend race at Darlington.

The second Richmond race of the year, in September, is the 26th and final "regular season" race. At the end of the Richmond race, the playoff field—the 16 drivers who will be eligible for the championship—is set.

The final ten races of the year visit eight tracks for the second time. The lone two single racetracks are Chicagoland (the first race in the Chase) and the final race of the season, where the champion is crowned at Homestead-Miami.

The Big Races

The biggest race to a NASCAR fan would probably be the one that's happening that week. That's because a NASCAR race in one of the three top-tier series is usually a pretty big deal wherever it's being held, but there are races in the Cup series that stand above the rest.

These "Crown Jewel" races are either prestigious, historic, or have large purses (the amount of money spread out among the field depending on where they finish). NASCAR fans and those inside the industry, including drivers, would probably argue about such a "Crown Jewel" list, but the five races that would generally be on anyone's list include:

1. The Daytona 500 (historic venue, first race of the season, largest purse)
2. Coca-Cola 600 (longest race of the season held at Charlotte, which many consider a "home track")
3. Brickyard 400 (historic venue, Indianapolis Motor Speedway)
4. Southern 500 (historic venue, Darlington Raceway, and one of NASCAR's longest-held races)
5. The August night race at Bristol (see below)

The first time I stood atop victory lane at Bristol Motor Speedway for a night race, it took my breath away. The half-mile high-banked track in Tennessee is super fast and has been characterized as "Racing fighter jets in a gymnasium." When you add lights (the spring race at Bristol is held during the day), the August night race at Bristol is nothing short of magical. For many years, a ticket to this very special event was something coveted by fans. There are legends that some fans would even will their tickets to their kids. Bristol Motor Speedway is the closest thing NASCAR has to a coliseum, and attending a race there, or taking time to watch it on TV, is necessary for a NASCAR fan.

The Daytona 500 is popular due to its being the first race of the season and winning there is a highlight on any racer's résumé. Known as the "Great American Race" or NASCAR's "Super Bowl," the Daytona 500 is usually the most watched on TV and the most anticipated after NASCAR's short winter break.

The Coca-Cola 600 (once known as the World 600) is held during the Memorial Day weekend and is NASCAR's longest race. The 600 is part of a smorgasbord for motorsports fans. It normally caps off a day of big-time auto racing starting with the Formula 1 Grand Prix in

Monaco, followed by the Indianapolis 500, with NASCAR getting all the attention at the end of the day as the race starts in the early evening and finishes under the lights.

The Brickyard 400 (which usually has a sponsored name each season) is held at the historic Indianapolis Speedway, and winners there all agree that a Brickyard win is one of the biggest they ever earn. By the way, if you ever see an IndyCar team "kissing the bricks" (the yard of bricks from the original surface that designate the start/finish line), after winning the Indy 500, they are doing a ritual started by NASCAR. In 1996, after winning the Brickyard 400, Dale Jarrett lined up with his winning team across the line of bricks, and they all leaned down and kissed them. The practice carried over to the Indy 500 and continues to this day.

Darlington Raceway in South Carolina is one of NASCAR's oldest tracks and its first paved superspeedway. The track held its first race in 1950 and has been holding races there ever since (only Martinsville Speedway has been on the schedule longer, holding its first NASCAR event in 1948). Darlington has a unique 1.366-mile oval layout. When Harold Brasington first built the track, it was on 70 acres of farmland he had purchased. The only caveat the former owner had was that a minnow pond on the west end of the property not be disturbed.[6] Brasington complied, and the result was an egg-shaped oval with the corners on the east end flatter and wider than the two on the west side, which are shorter and banked higher to accommodate the original landowner's request. Due to the notoriously rough surface of the track, Darlington has been nicknamed "The Track Too Tough to Tame." It is also known as the "Lady in Black" due to the propensity of drivers getting too close to the outside wall, scraping along it, and leaving a black mark and earning what's known as a "Darlington Stripe."

Who Do I Cheer For?

What makes NASCAR more interesting is not just understanding what is going on, but also having someone to cheer for. If you find a driver or team to root for, NASCAR becomes more than just a weekly race; it becomes an all-week affair. Focusing your attention on a single driver will open up more than just who is leading, or who won; it will allow you to see more than just a pack of race cars going around in a circle.

There is a great deal more that goes on during a race than just one car leading. There are races for position all over the track, and drivers who are fighting to make their way to the front, and drivers looking to win one of the first two stages. There are also drivers contesting the "Lucky Dog" (that is, the first driver one lap down) position, trying to stay ahead of the leader to avoid going a lap down, and drivers simply trying to finish inside the top 10.

In fact, there are many stories playing out during a race, and being able to follow these is key to really enjoying the experience. Being a fan of a certain driver allows you to not only see a story play out during a race, but also lets you follow that driver during the week, and through-out their career. Most drivers are active on social media now, allowing you to peer inside the life of a driver away from the track.

In the formative days of NASCAR, drivers who were stars usually never made much news away from the track. Today, however, nearly everything a driver does on and off the track is publicized. From how they did in a particular race, to signing a new sponsor, to getting married, having a child, or even what they did on vacation, a driver's life is much more exposed today, making NASCAR a much more immersive experience than ever before.

Trying to follow every driver, though, would be nearly impossible. So how do you figure out who to follow? There has not been a scientific study that reveals why a person becomes a fan of a particular driver. Asking fans why they root for a certain driver will yield answers that range from brand loyalty, either to the car make (e.g., Chevrolet, Ford, Toyota) or to a driver's sponsor (e.g., Lowe's, Home Depot); to the colors used in the primary paint scheme of the car; to where a driver is from, be it the fan's home state, or just a locale that a particular fan likes; to the physical attractiveness of a driver (driver Kasey Kahne, who races for Hendrick Motorsports, has a legion of young female fans; he may not win many races, but he remains one of the most popular drivers in NASCAR, and many female fans are quick to tell you why).

There are also a great many fans who became fond of a certain driver simply because their spouse or significant other cheers for them. Another way fans gravitate to a certain driver is through their workplace. For example, FedEx is currently the primary sponsor for driver Denny Hamlin, and many of his fans are employees of FedEx. The same with Lowe's Home Improvement, which has sponsored driver Jimmie Johnson for years. Lowe's gives back to its employees, as do several other NASCAR sponsors, with events that bring their driver to the workplace, or with organized trips to a race.

In fact, many times when Jimmie Johnson is interviewed, not only does he thank his crew (as most drivers do), but he also thanks the employees at Lowe's for their support, as well.

There are also what I call "bandwagon" fans who cheer for a driver just because everyone else does. Dale Earnhardt Jr. is currently NASCAR's most popular driver. His fans were built on the foundation of his father early on, but as his popularity grew, many fans followed him simply because others were doing so. The more fan merchandise, hats, t-shirts, and the like that is seen at the track, the more fans latch onto Earnhardt Junior's coattails. Of course, Dale Junior's easygoing demeanor and approachability don't hurt.

Anatomy of a NASCAR Race

Now that you know who you are going to cheer for, just what the heck is going on during an actual race?

Currently NASCAR Cup races range from 300–600 miles in length. The size of the track will dictate how many laps there will be to reach the advertised distance. The Daytona 500, for instance, is 500 miles total, on a 2.5-mile track, which equals 200 laps. When you see the name of a race, you can usually see how many miles or laps will be raced. Sometimes the number in a race name will be the number of laps, but miles are most often the number advertised. Sponsors will pay big bucks for the naming rights to a race, and with very few exceptions, the distance or number of laps will always be part of the name.

The Federated Auto Parts 400 at Richmond International Raceway will be 400 laps. Richmond is three-quarters of a mile long, so trying to run 400 miles at Richmond would equal over 500 laps, which could take quite a while. The Pure Michigan 400, on the other hand, is 400 miles long on the 2-mile Michigan International Speedway. Those 400 miles at Michigan equals 200 laps. The Goody's Fast Relief 500 at Martinsville Speedway is 500 laps on the .526-mile track.

No matter the track, starting in 2017, there are three stages of racing. The first two stages are spaced out, with the second stage near the halfway point of the race. The top 10 finishers in each stage earn championship points (10–1), and the winner of each of the first two stages gets one playoff point; the race winner gets five playoff points.

So there you are on your couch, favorite chair, or in the grandstands. Now what?

When you are watching an actual race, a great deal has already happened prior to the green flag. Cup teams normally arrive a day or two prior to the race. There have been practice sessions and a qualifying race (and in the case of the Daytona 500, two qualifying races).

Cars line up to practice at Charlotte Motor Speedway. Photo by Greg Engle

The final practice session on a race weekend is known as "Happy Hour" and actually lasts longer than an hour. It's known as "Happy Hour" because a team better be "Happy" with the car at the end of the session, since that's the last practice they will have prior to the actual race. The only other on-track time a team might have beyond practice is during qualifying, but NASCAR tightly controls what teams can do during qualifying, so most of the final work on the car must be done during "Happy Hour."

There are two types of schedules seen during a Cup race weekend based on whether an event is an "impound" or "non-impound" race. An impound race means that after qualifying, the race cars are locked down and can't have any major adjustments made prior to the race. A non-impound race will have more practice and a Happy Hour after qualifying, and teams can make adjustments prior to the race.

Teams can work on a car during an impound race weekend and are sometimes forced to. In this case, they are allowed to work on the car; however, they will be forced to start at the rear of the field on race day.

The biggest event prior to a Cup event on a race weekend is qualifying. What was once a lackluster affair changed in 2014. There was a time when NASCAR qualifying consisted of a single car making one- or two-lap timed runs with the starting order determined fastest to slowest, with the fastest driver being awarded the pole. The Cup field

Cars head to the track for "Happy Hour" at Martinsville. *Photo by Greg Engle*

was set this way for many years, with drivers who were not fast enough to make the field allowed a "provisional" starting spot if they were a past champion.

For the 2014 season, NASCAR instituted a new knockout style of open qualifying. NASCAR had implemented an open group qualifying procedure at the two-road course races in 2013, but adding it everywhere for 2014 was a game changer. The new knockout-style qualifying procedure was debuted at every track in 2014, with the exception of the Daytona 500 and the All-Star race (those procedures are outlined later).

Cars lined up on pit road prior to a knockout-style qualifying session. *Photo by Greg Engle*

It works like this: At tracks shorter than 1.25 miles in length, there are two timed rounds (30–10 minutes); at the tracks longer than 1.25 miles, there are three timed (25–10–5) rounds. For two rounds only, the top 12 cars move to the final round and compete for the pole. For three rounds, the top 24 move to the second round and the top 12 from the second round move to the final five-minute round. The final 12 compete for the pole and the top 11 starting spots behind. There is a ten-minute break between the sessions in a two-round qualifying format and a five-minute break between the three rounds in the three-round qualifying format.

The new format is now in use in all three top series and has brought a new level of excitement outside that of the actual races. If qualifying rains out, then NASCAR will set the field according to owners' points.

The only races that have a different qualifying format are the Daytona 500 and the annual All-Star Race.

For Daytona, the qualifying session is held on the Sunday of the week prior to the 500 and only sets the first two positions: pole and second place. On the Thursday prior to the 500, twin 150-mile qualifying races are held. These set the order of the field behind the front row. The Daytona 500 pole winner leads one of the twin qualifying races to the green, the second place starter the other. The race lineups are set by the qualifying speeds from Sunday.

Just like the format for the All-Star Race changes on a regular basis, so too does the qualifying format. Currently, the driver runs three laps, one that must include a four-tire pit stop; however, the pit road speed limit is lifted so cars can go as fast as they want there. The time of all three laps, including the pit stop, are then used to set the field.

The race cars have also undergone several inspections by NASCAR prior to the green flag. They are inspected prior to and after qualifying, and prior to and after the race itself. The top two finishers each week, along with a car chosen at random, are sent back to NASCAR's Research and Development Center in Concord, North Carolina, for further inspection that includes a complete teardown.

As you can see, when the green flag waves, a lot has already happened.

The field for a Cup race uses a rolling start; it comes after two or three slow laps behind a pace car, which is a car that does not race and many times is simply a regular passenger car (or pickup truck) fitted with special seat belts, a light bar, and flashing lights and decked out for a race sponsor or the track. When the lights are on, the pace car is in control of the field.

A NASCAR inspection station. *Photo by Greg Engle*

The only time you will see a pace car's lights not on is when it crosses the start-finish line. The fact that the lights turned off signifies that the next time around, the field will go green, either to start the race or to end a caution. The pace car will lead the field out of turn 4, then dive into the pits and give control to the starter, a NASCAR official, in the flag stand. The field remains at pace car speed until they reach the "restart zone" a designated area nearing the start-finish line, when the leader speeds up, the green flag is waved from the flag stand, and the racing is underway. The leader can't speed up too soon, nor can there be any passing behind the leader until that leader has crossed the line. "Jumping the restart" is grounds for a penalty, and you will hear sometimes that NASCAR is "reviewing the start." It's almost like a "false start" or an "offside" call in football.

The first few laps behind the pace car are called, not surprisingly, pace laps. The field is lined up two by two, with the pole winner starting at the front of the field on either the high or low side of the track, its driver's choice and will depend on the track. At some tracks, it's easier to pass on the outside, others the inside, depending on what's known as the groove.

All tracks have a racing "groove," or a certain area that is the fastest way around a track. The groove is not always the fastest, and there may be only one groove. There may be a "high groove" closer to the outside wall, or a "low groove" near the inside or apron of a track. The groove that is fastest is something drivers search for, or know from experience. Usually you can see the preferred groove at a track by the darker color of the pavement caused by rubber from the tires put down during practices and support races. The groove can, and oftentimes does, change with weather conditions or as a race progresses.

Once the green waves, and the field crosses the start-finish line for the first time, the race is under "green" conditions, the track is "hot," and the racing is underway. That racing will continue until cars need to be serviced for fuel, in which case they will pit under green flag conditions, or if there is a yellow flag displayed, a "caution" period begins and the cars are slowed. No passing is allowed on the track, and the pace car is in charge of the field. During a caution, once the pits are opened, cars will then pit. The order they leave the pits determines where they will line up on the track, which is why you will sometimes hear the phrase "the race off pit road."

Caution flags can be called for because of crashes, rain or another weather condition that makes racing unsafe, debris on the track that makes it unsafe to race, a car that has mechanical problems and has either slowed or stopped on track, or a vehicle that loses an engine and leaves fluid on the track.

NASCAR sometimes will also call a "competition caution," which usually happens because it rained the night before or the morning of the race, washing all the rubber out of the groove; because some of the practices were rained out; or for another technical reason. A competition caution usually happens early in the race, most often within the first 30 laps, and is announced before the race. Prior to a competition caution, cars can't be fueled, and if a team elects, cars aren't actually required to pit during a competition caution.

Starting in 2017 under the new stage racing, a white flag will be displayed one lap prior to the end of a stage followed by a green-white checkered flag to designate the winner of the stage, followed by a yellow flag when the top 10 cars have crossed the line, signifying a competition caution.

If there is no caution and cars are in danger of running out of fuel, the field undergoes a "cycle" of green flag stops. Some teams may use a fuel strategy during a race with pit stops. One car may decide to "short pit," that is, pit for fuel long before they need to. This allows the team not only to fuel the car, but also to make adjustments, and put on fresh tires. Many times, short pitting will allow a driver to make up time on the leader and stay out while the rest of the field undergoes green flag stops, thus gaining positions on the track. This puts the short-pitting car on a alternative pit cycle from others and can be used to gain an advantage. However, the strategy can fail, especially if a caution comes out before the need for a green flag stops.

Whether under green or during a caution period, pit stops can help win or lose a race. Pit stops are actually crucial to driver's success and an important part of a NASCAR race today.

During NASCAR's formative years, races were short and race cars rarely needed fuel or service during a race. As the races lengthened and cars needed to be serviced, it was done by the mechanics who worked on the car. In the beginning, stops were done using old-fashioned jacks to raise the car up to change tires, and fuel was added using several large cans. This process could take several minutes. By the mid-1960s, however, floor jacks with rollers were being used, and the stops were becoming more choreographed. The time in the pits was cut down, and spots were gained on the track.

One of the earliest innovators of pit stops was the Wood Brothers team. They practiced pit stops away from the track, and their drivers were winning because of it. In 1965, the Wood Brothers were hired as the pit crew for Indy driver Jim Clark. Clark won his only Indy 500 that year in large part due to the fast pit stops provided by the Wood Brothers. The choreographed pit stop was introduced to the rest of the motorsports world, and today it is commonplace.

The first pit crew specially recruited for raceway pit road duties is generally accepted as the one hired by crew chief Ray Evernham. This legendary crew chief put together a team of athletes with the specific purpose of servicing the car raced by his new young driver Jeff Gordon in 1993. The Hendrick Motorsports crew became known as the Rainbow Warriors in reference to the multicolored paint scheme on Gordon's car. Their motto was "Refuse to Lose." Other organizations followed, and soon purpose-built pit crews were the norm.

Today in NASCAR, the purpose-built crew trains like the professional athletes they are, and in fact many are former college athletes whose only job during the week at the race shop is to work on physical training as well as pit stop practice. Today, a well-trained pit crew can change four tires, add a full tank of fuel, and make minor adjustments to a race car in less than 12 seconds. A good pit stop can help keep a leader at the front of the field and allow a car to gain positions on the track. A bad stop can lose a race.

There are six crew members who are allowed "over the wall," that is, inside the actual pit box on pit road, to service a race car. A full pit stop consists of changing four tires and filling up the car with fuel. The crew will also wipe the front grill clean of any debris and make minor adjustments to the car if needed. This is all accomplished in under 12 seconds.

There is a great deal of strategy involved in a pit stop, some of which is done to help the car handle better on the track, or shorten the pit stop and gain spots in the field. Fuel mileage, how many laps are left, how the tires are wearing, or other feedback from a driver are just a few of the elements that play into pit strategy. At a track where tires don't wear much, for example, a team may elect to only put two tires on one side, while others take four. This move results in a shorter pit stop and a gain in positions.

NASCAR SPRINT CUP SERIES
TIGHT VS. LOOSE
CONDITIONS AND ADJUSTMENTS

TIGHT: Also known as understeer. This occurs when the front wheels lose traction before the rear wheels. It causes the stock car to have trouble steering sharply and smoothly through the turns as the front end pushes toward the wall.

LOOSE: Also known as oversteer. This occurs when the rear tires of the stock car have trouble sticking in the corners. This causes the car to "fishtail" as the rear end swings outward while turning in the corners.

During a pit stop, one of the crewmen will sometimes add or subtract spring pressure by attaching a ratchet and manually rotating it one way or the other. This tightens or loosens the spring and brings the frame and trailing arm forward or away from each other, applying more or less pressure on the tire when the car goes into a turn. This is known as adding or subtracting wedge.

1. Ratchet inserted by crewman
2. Ratchet extension
3. Side window
4. Rear window
5. Screw jack
6. Chassis frame
7. Coil spring
8. Trailing arm
9. Trailing arm end
10. Goodyear Tire

Photo courtesy of NASCAR

A driver may report trouble with a "tight" or "loose" condition. A "tight" condition, or "understeer," means when the driver tries to steer the car it won't turn sharply, or quick enough. A "loose" condition, or "oversteer," happens when a driver steers and the back of the car feels as if it wants to break loose or swing out. While no driver wants a tight condition, some drivers, particularly those who learned to race on dirt tracks that are usually slick, like a car that is a little bit loose.

During a pit stop, the crew chief may tell the team to adjust the "wedge" of a car based on feedback from a driver. Wedge refers to the amount of pressure a jack bolt on top of the rear spring puts on the rear tires. A crew member will be seen putting a wrench through the rear window of the car during a pit stop and putting a "round" (a turn of the wrench) of wedge in, or taking a round of wedge out (a turn of the wrench the other way). This adjustment can tighten or loosen the car's handling because it actually helps to shift the car's weight balance around. A wedge adjustment is usually only done on one side during a stop because the weight will then be shifted and the handling of the car will be affected. Imagine a perfectly balanced car on four springs. All the weight is equal; however, if we put more pressure on one spring, the balance shifts. This minor balance shift can change the way a car handles.

Other adjustments that can be done include changing the air pressure in new tires prior to putting them on, or by adding or removing "spring rubbers" (rubber spacers placed between spring coils that make the springs respond differently). Although usually done prior to the race and only under extreme circumstances during a race, the degree of camber in the wheels can also be adjusted. Camber refers to how the tire is vertically aligned to a car. Most times, the front tires will have a certain degree of positive or negative camber. The degree refers to how much the top of the tire is tilted inwards (negative) or outwards (positive) in relation to the frame and body.

Another adjustment that can be made is done by the driver in the car. Drivers can now raise or lower the car's "track bar" during a race from the driver's seat, something that began in 2015. The track bar (old timers still call it the panhard bar) is a lateral metal bar that controls the axis of the rear tires and centers them under the car's body. The bar is connected to the car's frame on one side and the rear axle on the other side. A track bar adjustment can also tighten or loosen a race car's handling.

During a pit stop, five helmeted crew members go over the wall:

- The rear tire carrier carries the tire from behind pit wall and follows the rear tire changer to the far side (left) of the car.
- The rear tire changer carries a high-speed pneumatic air gun and removes the five lug nuts holding the tire on. The rear tire changer then removes the old tire, the rear tire carrier puts the new one on, and the tire changer tightens it down with the five lug nuts. The two then run to the other side of the car and repeat the process, if the stop is for four tires. The rear tire changer is often the crew member seen making the wedge adjustments with a wrench placed in a hole in the back window.
- The front tire carrier and front tire changer mirror the work being done at the rear of the car.
- The jackman carries the 20-pound hydraulic floor jack used to raise the car. Once all tires are tightened on one side, the jackman drops the car and runs around to repeat the process on the other side. When the car is dropped the second time, it is the signal to the driver that the work is done and they are free to leave the pits.
- The gasman is the crew member, usually one of the biggest persons on the team, who uses two 12-gallon fuel cans (each weighing 80-plus pounds) in succession, to dump fuel into the car's fuel cell.
- During a stop, you'll see other support crew members leaning over the wall rolling and catching tires, keeping the air

hose-free, and grabbing empty fuel cans. These crew members wear no helmets and must stay behind the wall.

- On rare occasions, NASCAR will allow an extra crew member over the wall, but only to clean the windshield or aid the driver if needed.

Not long ago, a NASCAR official could be seen going over the wall with the crew to watch the stops; in fact, a NASCAR official was assigned to watch two cars. That changed when NASCAR instituted a new officiating system that put fewer officials on pit road and now uses an extensive network of cameras that are monitored by officials in a trailer in the infield.

Pit road during a race is a dangerous place, and through the years accidents on pit road have claimed the lives of crew members (though thankfully none since the 1990s), and others have been injured during pit stops. That's why crew members wear fire suits and helmets, and why NASCAR has a pit road speed limit.

Not all pit road visits are welcomed ones. If a car is damaged during a race or suffers a tire failure, it must pit under green for repairs and will usually lose several laps. Also keep in mind that race cars don't have fuel gauges. How much fuel mileage a race car gets is a very educated guess, but even those can be wrong or a team and driver may try and stretch their fuel to the absolute limit.

A fuel mileage strategy can win a race if done right. However, if a car runs out during a race, a long slow roll to the pits will usually end their chances for a race win. NASCAR can also use the pits as part of a penalty. If a driver is penalized, for speeding on pit road for example, NASCAR can force them into the pits for a "pass through" penalty, meaning the car must make one pass on pit road, at pit road speed. Another example is a "stop and go" penalty, where the driver comes in, stops in their pit stall, then goes out again, all the while maintaining pit road speed. Either penalty is costly to a driver and team.

Other common pit road penalties in addition to speeding include entering the pit stall too soon (a driver can't drive through more than two stalls prior to their own), pitting outside the painted lines of the stall, crew members going over the wall too soon (prior to the stall before their cars), leaving with equipment still attached (e.g., a fuel can or a wrench still sticking out the back window), or an uncontrolled tire—meaning a tire carrier or the crew member designated to catch it lets the tire get away and roll outside the pit box. NASCAR will many times enforce a rule that says a leader cannot pass the pace car when pitting. This practice, called "pulling up to pit," happens during a yellow flag. The leader of the race will drop off the track to head to pit road. However, the driver cannot pass the pace car that is still on the track. If penalized, NASCAR will normally force a driver back to the pits and hold them for a lap.

NASCAR can also penalize a driver by sending them to the rear of the field or to the tail end of the longest line for such things as missing the prerace drivers meeting or driver introductions, or if the team makes adjustments to a car during an impound period.

A pit stop and the team's pit strategy is all directed by the crew member on top of the pit box, the crew chief. The crew chief is the coach, or the quarterback, the one who directs how the car will be set up and pitted; he interprets the feedback from the driver and directs the changes to be made. Some crew chiefs are nearly as famous as drivers are, and a winning crew chief can be just as valuable to a team as a winning driver.

Other members of a race team's crew include a car chief, who works with the crew chief to strategize car setups and make sure the needed changes are made. There's also an engineer who uses all sorts of performance data and measurements—including camber degrees and even tire pressures—to calculate how a car should be set up or adjusted.

There is another important crew member that needs mention here, and that is the "spotter." This is a crew member, many times a former driver, who is stationed high above the track, usually on top of the main grandstand area. The spotter's job is to relay to the driver via radio where that driver is located on the track, who is around them, and if they are clear of other cars. Their instructions are short: "Clear high," "Clear low," "One on your outside . . . still there . . . clear" are just a few examples of what you might hear.

A spotter can also tell the driver another line to take on track, perhaps one the leader is running a bit faster. On the large superspeedways and road courses, tracks where one spotter may not be able to see the whole track, a team will usually have more than one spotter. A driver can make very few moves during a race without the help of the spotter.

Besides pitting due to the fact that a car is low on fuel, why pit? Shouldn't the car's setup be good prior to a race? Adjustments during a race are needed because the conditions of the race typically change. Weather plays a big factor in a car's speed, and as the day gets hotter, a track will usually get slicker. In addition, teams may only practice during the daytime for a night race. Of course, as pointed out earlier, pit stops and adjustments can play into a race team's strategy.

There are plenty of other things going on during a race in addition to pit stops and pit strategy.

During a caution period, you might see a group of cars take off and race past the pace car. These are the "Wave Around" cars, those one lap down. They are allowed to make up a lap but must start at the rear of the field. They can't pit during the caution, only once the green comes back out. The "Wave Around" cars are in addition to the "Lucky Dog," which is the first car a lap down (unless the driver eligible for the Lucky Dog was responsible for the caution). The Lucky Dog gets an automatic lap back from NASCAR and can pit under the yellow, something called the "free pass."

All the rules about getting laps back, the Lucky Dog, Wave Arounds, and such stem from a policy that came about in 2003. Prior to that year, when a caution was displayed, those cars a lap down could continue to race to the start-finish line in order to gain a lap back. Many times, under a "gentleman's agreement," the leader would actually slow and allow cars to get a lap back. This practice, though, caused more

FREEZING THE FIELD

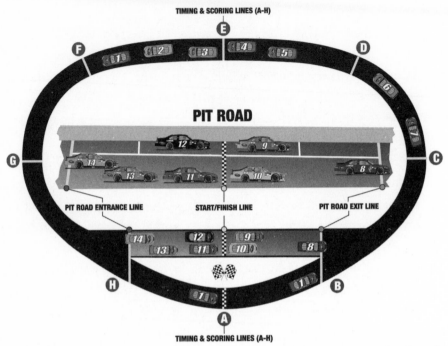

TIMING & SCORING LINES (A-H)

PIT ROAD

PIT ROAD ENTRANCE LINE   START/FINISH LINE   PIT ROAD EXIT LINE

TIMING & SCORING LINES (A-H)

In an effort to further ensure the competitors' safety, NASCAR announced in September of 2003 that racing back to the caution would no longer be permitted in the NASCAR Sprint Cup Series, NASCAR XFINITY Series and NASCAR Camping World Truck Series. That led NASCAR to institute a new procedure in which the field is "frozen" on the race track once the caution flag is issued.

The cars' positions are determined by the previous timing-and-scoring line they passed on the race track. On the backstretch, cares 1-3 are scored by their running order when they passed timing and scoring line E; cars 4 and 5 are scored by their positions when they passed timing and scoring line D; cars 6 and 7 are scored by their positions when they passed timing and scoring line C.

While the positions of the competitors on the race track will be "frozen," the pit lane, however, will remin active at this time with pit-road speed in effect.

The cars that are pitted from pit-road etrance to the start/finish scoring line that extended acorss pit road before the race leader (No. 1 red car approaching scoring line A)

reaches the same line on the race track. Should any of these cars on pit road reach that point first, they will not lose a lap to the leader. Should the leader reach scoring line A before cars 11-14, they would lose a lap to the leader. Those cars that are pitted from the start/finish line to the pit-road exit - cars 8, 9 and 10 - must reach the pit-road exit scoring line before the leader (No. 1 red car approaching scoring line B) to avoid going a lap down. Example: car 8 would not go a lap down, while cars 9 and 10 would.

Should a driver attempt to speed in pit lane to avoid going down a lap to the leader, that driver will lose a lap in addition to being moved to the tail end of the longest line. Should the race leader not slow immediately for the caution in an effort to put the pitted cars a lap down, the leader will be penalized by being sent to the tail end of the longest line and all pitted cars will retain their lap positions.

Source: NASCAR

NASCAR

than a few accidents through the years and prevented safety crews from going on the track and getting to the scene of an accident in an expeditious manner. Under the new rules, the field is frozen when the caution is displayed, and everyone slows. This allows safety crews to get to an accident scene faster.

There was a further change for the Wave Around and Lucky Dog drivers in 2009, regarding restarts. That year, NASCAR instituted the "double-file restart shootout-style" rule. The procedure first used in the All-Star race the previous year put those cars making up a lap behind the lead lap cars. Prior to this, the Wave Around and Lucky Dog cars restarted in a single file line next to the leader, who led the line of cars already on the lead lap. The new rule in 2009 eliminated that, and now the lead lap cars line up first, the second cars line up next to them, the third behind the leader, the fourth beside the third-place starter, and so on. Those a lap down line up behind the last car on the lead lap.

When the field is green, you may hear terms that sound odd. Such words include:

Drafting: Mainly a factor at the superspeedways and faster tracks, drafting refers to an aerodynamic condition. Obviously, race cars are fast, and that's mainly due to the aerodynamics, or airflow over the car. Normally, the clean air flows across the top of a race car and becomes turbulent on the backside. Drafting is a process of two race cars running close, nose to tail, that allows this air to flow over two or more cars, smoothly creating a vacuum with the turbulent air at the back of the last car in line. When cars are running in a draft, they are often seen very close, sometimes even touching the rear and front bumpers. Two cars drafting will always run faster than a single car, most often several miles an hour faster. That's why you see packs of cars racing together at Talladega and Daytona, and why a car that "loses the draft" or is "hung out" will drop back. The turbulent air behind a car is also called "dirty air." Not long ago, cars would run in a two-car draft. This practice, however, was outlawed in 2014, along with "bump drafting," where two cars would draft, with the rear car "bumping" the front one to gain speed. For a couple of years, these two-car partners were the norm at Talladega and Daytona.

Slingshot: A move first used famously by Junior Johnson in the 1960s, the "slingshot" is another move seen at superspeedways. Relying on the draft, the second car in line will use the vacuum created in the draft to gain momentum in the corner, then move out and pass as the cars exit the turn. This move has created many exciting finishes throughout the years, including the finish of the 2016 Daytona 500, when Denny Hamlin used a slingshot move to pass leader Martin Truex Jr. exiting the final turn and win the race by the closest margin in history. The slingshot is also why the start-finish line at Talladega is closer to turn 1. NASCAR founder "Big Bill" France Sr., who built the superspeedway, moved the finish line to give drivers an opportunity, for one final slingshot, to move on the final lap.

Restrictor plate: This is a thin metal plate that restricts the amount of airflow into the engine. It's only used at Talladega and Daytona. First mandated in 1988 after Bobby Allison crashed at Talladega the previous year and his car flew up into the fence, injuring several fans, the plate saps horsepower from an engine and slows down the cars at Talladega and Daytona. With the electronic fuel injection in use in the Cup series today, the plate is still used and placed between the throttle body and the engine.

Side Drafting: With the new low downforce package being used in the Cup series, which is supposed to put the emphasis on the driver and less on the car sticking to the track, side drafting has become more prevalent. Side drafting is similar to a slingshot and happens when the lead car in a draft loses speed and momentum as another car moves up from behind it. When used properly, the car passing will usually be able to pull ahead, while the car that is being passed gets "loose" and slows.

There are other things that can affect a race car's aerodynamics. If a driver reports that the "splitter" is hitting the track or damaged, what is being referred to is a body piece that runs along the bottom of the front of the car. Teams will try and have the splitter as close to the ground as possible, as it helps "seal" the car to the track, providing front down-force. However, many times a team will put a splitter so close to the ground that it will bottom out, causing anxious moments for drivers.

The splitter works with the "spoiler," which is a piece of metal, like a blade, attached to the rear deck lid of the race car. It serves to funnel the air down across the back of the car and help rear downforce. The more downforce a race car has, the more traction it will have, and the faster it will go.

In recent years, NASCAR has experimented with "low downforce" packages, in an effort to take away some of the downforce. A race car with high downforce not only goes faster, but is also a bit easier to drive, as it requires less input. NASCAR's aim in taking away some of this downforce is not to slow cars down, but to put more emphasis on the driver.

With the higher downforce, passing on track was tougher, and many races saw little change in positions. With the new lower downforce, passing has not only become easier, but many agree the overall racing has become more exciting.

Not all the slick moves are reserved for the superspeedways. On short tracks, drivers may use a move called a "bump and run" to pass. The move that many attribute to Dale Earnhardt Sr. will have the car behind bump the car in front while entering a turn. If done correctly, the lead car will lose traction and be forced up the track or get loose enough that the car behind can pass. If done incorrectly, the lead car will spin, sometimes taking the car behind with it.

Dale Earnhardt Sr. was famous for "putting a bumper" to someone. At the August 1999 race at Bristol, with only a few laps to go, Terry Labonte passed Earnhardt for the lead. As the duo raced on the final lap, Earnhardt hit Labonte's car hard entering turn 2. Labonte's car spun, and Earnhardt roared past him and onto victory. Earnhardt later said he didn't mean to wreck Labonte, only "rattle his cage." The two had a similar finish four years earlier, with Earnhardt again bumping Labonte coming out of turn 4. In that finish, however, Labonte's car crashed into the wall but was pushed by Earnhardt across the line sideways, giving Labonte the win.

The "bump and run" move has been used many times over the years, famously and infamously, by many drivers to win races.

No matter what happens on pit road, or what moves are done on the track, the race will eventually end and a winner—the driver shown the checkered flag at the start-finish line by the flag person—will be declared. The flag person is the NASCAR official who takes directions from those in control of the race from the track's tower and referred to as "race control."

Flags

Flags are used during a race to communicate with drivers. They are somewhat archaic, since NASCAR can communicate to drivers via radio, but they remain an important part of the race.

Here's a breakdown of the flags used in NASCAR:

Green flag: This starts the race, or restarts a race after a caution period. Another driver cannot pass the leader until the green flag is displayed, and as long as it is displayed, the track is green and racing can continue.

Yellow flag: This flag comes out during a caution period and tells the drivers to slow down. The field is frozen, and no passing can occur. The yellow flag is waved most often to signify a crash, weather, debris on the track, or another condition that makes racing unsafe.

Red flag: When this flag is waved, everything must stop at once. Many times, the red will be shown to end practice, when rain makes racing impossible, or when a particularly bad crash has occurred. The red flag is usually preceded by the yellow flag. Once the pace car has the field under control, they are stopped on pit road, or another part of the track, and then the red is shown. Under a red flag, no one is allowed to work on a car for any reason anywhere on the track, including pit road or the garage.

Black flag: When a driver is shown the black flag, the driver must respond, usually meaning they will have to pit for some reason. NASCAR has often penalized the car for speeding on pit road, or some other infraction, and the driver is being told to pit for either a drive through or stop and go penalty.

On other occasions, the black flag is displayed to a car because of some type of damage to the car, and that car is putting debris on the track, or they are not maintaining minimum speed. Most often, a car that has been damaged earlier but has subsequently been repaired and returns to the track will cause this, as the repairs may not have been done well enough to allow the car to maintain a minimum speed (expressed by NASCAR prior to the race as the time it takes to complete a lap, not as miles per hour).

Why would a driver who has crashed want to return to a race where there is no chance of winning? Points: the season point standings can be so close that a few markers either way can make a difference. Every lap a driver can run could make the difference in the final results at the end of the race. Even one position gained can mean another point. If a driver can return to a race, he can make up spots. For example, if other cars drop out (scored with a "DNF" for "Did Not Finish"), that is a good opportunity to make up ground.

Black flag with diagonal white stripe: This is another flag a driver never wants to see. Once a driver is given the black flag, they must respond. If after five laps, that driver fails to respond, they are shown the black flag with the white diagonal stripe. This means that NAS-CAR has stopped scoring the driver: the driver's laps are not counting and whatever position they are in will be the one they finish with, if they don't respond. While some drivers may get close to the five-lap limit while their crew chief argues about an infraction with NASCAR officials, none will usually go beyond the five-lap limit.

Blue flag with diagonal yellow stripe: This is also known as the "move-over" flag and is shown to cars that are one lap or more down as the leaders approach them. The flag tells the slower drivers to give way.

White flag: The white flag is waved at the leader as that driver begins the final lap of the race. Under NASCAR's green-white-checkered, or overtime rule, once the white flag is displayed, the next flag ends the race, whether that be a yellow or the checkered.

Under the green-white-checkered rule, if a yellow flies prior to the white flag, the field will be given the opportunity to finish the race under green, even if the laps extend beyond the scheduled distance. With the green-white-checkered rule, the field is given the green flag, then the white, then the checkered.

If, however, a caution happens after the white flag has been waved, the field is frozen, and the race is over.

Checkered flag: This is the flag that every driver wants to see. It declares the winner as the lead driver crosses the start-finish line. Often, the race winner is given the checkered flag after the race, and that driver will parade around the track with the flag waving.

Green-white-checkered flag: Introduced in 2017, this flag designates the end of a stage and is shown to the winner of that stage.

Other notes about flags: If you see the NASCAR official in the flag stand hold up two flags that are crossed, this indicates that the race has reached the halfway point. In addition, most tracks will have lights that will display green and yellow when needed. The pits too have lights, green and red, along with a NASCAR official at the opening of pit road who will display a green flag when the pits are open or a red flag with a yellow X to show the pits are closed. This is all controlled by officials in race control in the tower.

What's It Like to Be a Driver?

Though it's important to understand the fundamentals, until you hear firsthand what it's like for a driver to race 400, 500, and, for one race a season, 600 miles, you may not believe all that goes into it.

Here's what it's like to race those many miles straight from a driver.

RACING FLAGS

NASCAR officials help signal messages to drivers during races by waving an assortment of colored flags. The flagman, who is always located on a stand high above the start/finish line, plays an important role during an event.

GREEN FLAG – Displayed at the start of the race and also for restarts during the race. Cars must maintain position until they have crossed the start/finish line. The polesitter at the race start – and race leader on restarts – controls the pace and cannot be passed prior to the green flag waving.

YELLOW FLAG – Signifies caution and is given to the first car passing the starter immediately following the incident that caused the display of the flag. All cars must slow down immediately to a pre-determined pace and hold their position behind the pace car.

RED FLAG – Signifies the race must be stopped immediately, regardless of the position of the cars on the track. The red flag shall be used if NASCAR officials decide the race should be stopped, usually for safety and/or competition-related reasons. Cars will be brought to a stop in an area desig-nated by NASCAR officials. Repairs or service of any nature or refueling, whether on pit road or the garage, will not be permitted when the race is halted due to a red flag, unless the car has withdrawn from the event.

BLUE FLAG WITH DIAGONAL YELLOW STRIPE – Although this flag is displayed regularly during most during races, it is probably the least recognized. This flag is displayed to drivers, who are a lap down or significantly slower, that are about to be passed by lead-lap cars. Drivers who are shown this flag must yield to the faster lead-lap cars.

BLACK FLAG – Display of this flag requests that the car go immediately to the pits and report to the NASCAR official at the car's pit area. The car can receive a black flag for a variety of reasons, including a driver/team infraction, or a potential problem with the car reported by NASCAR officials that warrants a closer inspection in the pits. It does not mean automatic disqualification.

BLACK FLAG WITH DIAGONAL WHITE STRIPE – At the discretion of NASCAR officials, if the driver does not obey the black-flag directive, the driver may then be given the black flag with a white cross at the start/finish line to inform the driver that any additional scoring of the car will be discontinued until further notice.

WHITE FLAG – Waves when the driver in the lead begins the final lap of the race.

CHECKERED FLAG – The most famous of all flags, the black and white checkered flag is displayed when the winner has crossed the finish line. All cars on the track will take the checkered flag once.

Source: NASCAR

NASCAR

Kyle Busch is the 2015 NASCAR Cup champion. However, his title didn't come easy. Busch often races in the support races in the Xfinity and Truck series in addition to Cup races. In February of 2015, Busch was racing in the season-opening Xfinity race at Daytona International Speedway when he was involved in a horrific crash that left him with a broken right leg and broken left foot.[4] Busch would undergo several surgeries, extensive physical therapy, and miss the first eleven Cup races of 2015. He would return, be granted a waiver by NASCAR, and win four races, including the finale at Homestead to win the title.

Prior to the 2016 Brickyard 400 at Indianapolis Motor Speedway, Busch explained that it's difficult for fans to understand just what drivers go through.

"I think it's hard for people to understand what exactly we feel throughout an event," Busch said. "I've heard a lot of people over the years that have gone and done the Richard Petty Driving Experience or the riding experience and they've certainly got a taste of what we do on a weekly basis and their quotes to me are, 'Man, we've got a heck of a lot more respect for you. That was a lot more than we anticipated or expected that we'd feel.' So, you always get that.

"I always encourage people to go get a chance to get in a race car to kind of feel and see what we do," he added. "To do it for four, four and a half hours on every single weekend it does take a toll on your body. I can say that now because when I was eighteen, nineteen, or twenty years old or younger I really didn't feel it. I could go through a whole season and I really didn't feel it a whole lot. Well, now I'm thirty-one and I feel it a heck of a lot more.

"Certainly, I remember it late last year getting towards the end of the year I actually still felt pretty good—I only ran half of the year," Busch said. "This year now I've run from the beginning of the year, I'm getting to about the halfway point, and I'm feeling the same way I did at Homestead and we still have another half of the year to go. So, certainly you've got to modulate your body and take care of your body the best you can. I feel like there's a lot of things you can do off the track that can help that and I try to do all of that stuff as much as I can."

A Final Thought on NASCAR's Rules

Trying to list all of the intricate rules involved in NASCAR would require a book of its own. The NASCAR rulebook is indeed a large one and covers every conceivable area of racing. It is so large, in fact, that it is now completely online.

There are rules that cover everything from the size, composition, and placement of every part on a race car to how members of NASCAR are expected to behave. Race and inspection procedures are spelled out in detail, along with penalty and appeal procedures.

Those penalty and appeal procedures were overhauled in the past few years and are now clearer than ever before. Penalties are broken into six classes. Minor technical infractions are in the P1 class and can lead to lost track time and other relatively light punishments; violations affecting the internal workings and performance of the engine fall under P6, on the other hand, and could lead to the loss of 150 points, a fine of at least $150,000, and suspensions.[5]

A team can receive the lowest penalty, a written warning, with every fourth one earning the team a loss of pit stall selection, something that can be very valuable during a race. After the fourth warning, the slate is wiped clean and starts over. The most common reasons for a written warning are failing prerace inspection. If a team's car has issues during a prerace inspection, NASCAR will allow them to correct the mistake and go through inspection again. This usually results in a written warning.

Another common infraction, especially starting in 2016, was the violation of the "lug nut rule." In 2015, when NASCAR changed the way they officiated on pit road, the rule that all five lug nuts had to be on a wheel was no longer enforced, but instead left up to teams. When NASCAR officials were physically present at each pit stop, that official could see if all five lug nuts were being put on, and if not, that car was brought back in. That ended in 2015, and teams found that they could save time on pit road by putting on fewer than five lug nuts to hold on a wheel. While it worked well for some, other drivers were uncomfortable, and more than one driver lost a chance for a race win by being forced to pit under green with a loose wheel.

In April of 2016, NASCAR began enforcing the five lug nut per tire rule again. They did this by checking the wheels postrace. The penalty was considered a P2 level, and several crew chiefs in all three series were fined and suspended for a race.

The other big penalties that fans will most often see include a car that fails postrace inspection. This usually happens when a car is found to be too low in postrace inspection. When a car fails, many times it is taken back to the R&D center for further inspection. The penalty that results is usually a fine and points taken away from the driver and team owner. Crew chiefs, drivers, and others can also be put on probation, meaning that another infraction of the same type could lead to even bigger penalties.

The biggest penalties in the sport occur when a team is caught altering a car in some way. NASCAR crew chiefs are famous for pushing the envelope when it comes to trying to get more speed from the cars. Many times, if NASCAR inspectors find an issue like this prior to the race, the team is forced to correct it—no harm, no foul. However, if the infraction is found postrace, the penalties can be, and usually are, severe.

The biggest fine in NASCAR history is still the $300,000 that was handed out to the Michael Waltrip Racing team after a 2013 incident at Richmond. The penalty also carried a 50-point deduction that knocked driver Martin Truex Jr. out of the Chase, giving the spot to Ryan Newman.

There have been plenty of NASCAR penalties through the years; here are a few recent examples:

In 2015, NASCAR caught several teams trying to manipulate tires, in what became known as "Tiregate." The speculation was that teams were putting small pinholes in the tires so they would slowly bleed out air as they heated up, keeping the tire inflation the same over the course of a race run. NASCAR began seizing random tires from teams after races and in April issued a P5 penalty to the Richard Childress Racing team of driver Ryan Newman. His crew chief was fined $125,000 and suspended for six races. Two other crew members, a tire specialist and the team's engineer, were also given six-race suspensions. Both the driver and team owner were docked 75 points. No one in the garage was caught manipulating tires again.

Driver Greg Biffle's crew chief was suspended for several races in 2016 after NASCAR officials said they found in postrace inspection that the team presented a car with a body that didn't meet NASCAR specifications, or that had not been approved by NASCAR prior to the race.

Boys, Have At It

There are also penalties that fall under the "behavioral" category, referring to something a driver or crew member does. These usually involve negative emotions, and while they most often occur outside the car, there have been times when a driver's emotions have gotten the better of him and a car has been used to indicate his displeasure.

There is a sort of unofficial policy when it comes to how NASCAR polices behavior. Several years ago, NASCAR fans accused the sport of being too tough on the way drivers settled disputes. In 1979, the fight between the Allisons and Cale Yarborough at Daytona had helped bring NASCAR into the mainstream. However, with big sponsors paying big money, sponsors wanted drivers who could not only race, but conduct themselves in a manner that befitted a brand. By the early 2000s, drivers were polished and proper. The old days of hard-partying drivers nursing hangovers and settling scores off the track with fists seemed over.

Fans complained that the personalities drivers once had were stifled by corporate America. The drivers, many said, were like robots constantly spewing advertisements for their sponsors. NASCAR finally reacted in 2010, announcing during the preseason media tour in Charlotte a new philosophy characterized by the phrase "Boys, Have At It," said by the NASCAR vice president of competition at the time, Robin Pemberton. The policy meant drivers could open up a bit more on the track, but there were limits as to the extent.

In 2011, during a Truck race at Texas Motor Speedway, Kyle Busch, who was known to have a hot temper, deliberately wrecked driver Ron Hornaday in retaliation for an earlier incident involving the two. Hornaday had been a contender for the Truck series title until that crash, and subsequent poor finish.

NASCAR wasted no time taking action. The Truck race had been on Friday at Texas Motor Speedway. Busch, who almost always races all the support races during a weekend, was banned from racing on Saturday and Sunday at Texas, after NASCAR declared Busch's actions to be "detrimental to NASCAR." He was also later fined $50,000.

Kyle Busch wasn't the first driver to be "parked," as the term for suspension from races is called.

In 2007, driver Robby Gordon (no relation to four-time champion Jeff) was parked at Pocono for an incident at an Xfinity Series race in Canada the day before. The race leader, Marcos Ambrose, spun Gordon, who was second, during a restart. Another caution came out, and Gordon charged through the field and intentionally spun the leader in retaliation. Gordon felt he should be the leader, but NASCAR said he was supposed to restart in 13th position. Gordon stayed in the top spot on the restart, and NASCAR officials "black flagged," meaning the black flag was displayed to Gordon, ordering him in to the pits. Gordon ignored the black flag and continued racing.

When a driver is "black-flagged," he must respond by pitting. If he doesn't do so within a few laps, NASCAR will stop scoring that driver, and that's what happened with Gordon. Kevin Harvick was the official winner of the race; however, Gordon was celebrating as well, insisting he had won. NASCAR parked Robby Gordon for the Cup race at Pocono the next day.

Harvick, meanwhile, knew all about NASCAR "parking" drivers, since he himself had been parked in 2002. Harvick, also known for his fiery temper, was competing in the Truck race at Martinsville Speedway on a Saturday, a day prior to the Cup race. He tangled with driver Coy Gibbs several times, eventually drawing a warning from NASCAR. Harvick ignored the warning and crashed Gibbs out of the race. NASCAR black-flagged Harvick, parked him for the race, and subsequently parked him for the Cup race the next day.

Most of the penalties for "actions detrimental to NASCAR" are for physical tangles that happen between drivers out of the cars, or between crew members who often face off in the garage or on pit road when their respective driver is involved in an on-track incident. Under the "Boys, Have At It" policy, pushing, shoving, finger-pointing, and yelling are tolerated. When fists begin to fly, however, NASCAR will usually be quick to step in. Once punches are thrown, NASCAR will usually put an end to it, and heavy penalties are usually levied.

Kyle Busch isn't the only member of the Busch family who races in NASCAR and is known for having a quick temper. Older brother Kurt is a 2004 Cup champion and has been involved in many incidents, several of which occurred on the track, but many more that took place off the track.

Kurt Busch was racing for one of the top teams in NASCAR, Team Penske, in 2005. With two races to go in the season, he was eighth in the Chase standings. On the Friday prior to the Cup race at Phoenix, Busch was arrested for suspicion of DUI by the Maricopa County Sheriff's Office. NASCAR took no action, but the team suspended Busch for the final two races of the season. On the legal front, Busch was ordered to perform fifty hours of community service. Ironically, when the Cup series returned a year later, Busch was made an honorary Maricopa County Sheriff's Deputy.

In 2011, Busch reportedly confronted a media member on pit road after an on-track incident, then later in the media center, he was asked about his comments. Busch denied he had said anything and when showed the written transcript snatched it from the reporter's hand, tore it up, and walked out of the media center.

Although publicly NASCAR took no action for the incident, it had been well known that NASCAR had been privately fining drivers for "actions detrimental to NASCAR," which included drivers making disparaging remarks or actions. The practice of privately fining, drivers would end after an uproar from the media, but even today, drivers will still rack up an occasional fine for making disparaging remarks, especially on social media.

The incidents at Richmond involving Kurt Busch and the media in 2011 weren't the only ones to happen that year; in fact, reports of such incidents only escalated as time progressed. At the final race at Homestead in November, Busch was forced to retire from the race early due to mechanical issues. ESPN reporter Jerry Punch wanted to interview Busch live. The two were forced to wait until the appropriate moment to go live. While waiting, Busch began to berate Punch in a profanity-laced outburst captured by a fan nearby and later posted on the Internet. NASCAR fined Busch $50,000, and Team Penske issued an apology. Soon after, Busch and Team Penske parted ways, although publically the split was called "mutual."

Kurt Busch's issues continued into 2015. Toward the end of 2014, news reports began to surface of a domestic violence investigation involving Kurt and his longtime girlfriend, Patricia Driscoll. The couple had broken up earlier in the year, and Driscoll alleged that Kurt had assaulted her during the race weekend at Dover in September. The allegations lingered into 2015 as the investigation continued. Only days prior to the Daytona 500, Kurt was suspended by NASCAR for the allegations. There was a hasty appeal by Kurt, which he lost. Charges were never brought against the driver, but Brian France insisted that the indefinite suspension would not be lifted until Kurt completed NASCAR's reinstatement program. Kurt would complete the program, and the suspension lifted in March.

Of all of the NASCAR drivers known for having a temper, however, Tony Stewart has seemingly led the way. The three-time champion has been fined by NASCAR on several occasions for fighting with other drivers, and once for even pushing a photographer after a race in 2004. That incident led to Stewart's sponsor at the time, Home Depot, taking the unprecedented step of fining him. The $50,000 from that fine went to charities in Stewart's hometown.

Stewart mellowed in the years leading up to his retirement after the 2016 season. While he's never been suspended by NASCAR, Stewart has missed Cup races.

Feuds between drivers are as old as the sport. Petty-Pearson and Earnhardt-Gordon are two examples of duos with a history of feuding on the track. Nevertheless, these racers respected each other away from the track. The "Boys, Have At It" policy allowed the Bad Boys to find where the line was drawn, but it also helped foster several feuds between drivers.

In one of the first incidents under the "Boys, Have At It" policy, Brad Keselowski had an ongoing feud Carl Edwards that ended with Keselowski 's car being sent into the fence upside down at Atlanta in 2010. Edwards was parked for the rest of the Atlanta race and later was given only probation by NASCAR.

No driver had been suspended by NASCAR for on-track antics since Kyle Busch in 2011. That was until 2015, when a feud between Joey Logano and Matt Kenseth reached a boiling point.

Under NASCAR's elimination-style Chase, drivers are eliminated after three races in every round. At the fall race at Kansas Speedway, Kenseth, facing elimination from the Chase, was leading with only a few laps to go. A win would assure Kenseth of moving to the next round in the Chase. Logano, running second, caught Kenseth with four laps to go and, using a bump and run, passed Kenseth, who was spun in the process. Logano won, and Kenseth finished outside the top 10. NASCAR termed it a racing incident and took no action. Kenseth would be eliminated from the Chase the following week.

The action, however, would come back to haunt both drivers.

Two races later, at Martinsville, Logano was the class of the field. He led the most laps and was leading the race nearing the end, when Kenseth, who had crashed with Logano's teammate Brad Keselowski earlier in the race, slowed, then deliberately charged up into Logano, pushing him into the wall. Logano lost the race, and his chances at winning the championship.

NASCAR suspended Kenseth for two races. Kenseth later said he had no regrets.

Also in the last decade, NASCAR took actions to institute a new substance abuse policy, starting in the 2009 season. It contained a comprehensive list of banned substances, along with reasons the sanctioning body could use to require a member to test, and how samples would be handled. Also new was the use of random sampling of members. A computer randomly selects members and requires them to provide a sample for testing, usually a urine sample, within a prescribed period.

If a member of NASCAR is suspended for failing a substance abuse test, they can earn their way back into the sport through NASCAR's Road to Recovery program. The program uses counseling, education, and more testing, and once a member completes the program, that person can, and usually will, be reinstated.

The new policy got its first real test, and became even more detailed, in its first year. On May 9, 2009, NASCAR announced that longtime driver Jeremy Mayfield had tested positive for a banned substance. He became the first driver to be given an indefinite ban under the new substance abuse policy. Ironically, in February, a crew member on Mayfield's team had also tested positive and given an indefinite ban.

Mayfield released a statement shortly after that indicated the crew member would take part in the Road to Recovery program, or face termination. When Mayfield himself tested positive, however, he chose a different route.

Mayfield denied he had taken anything other than over-the-counter medications and a prescribed drug for an attention deficit disorder. One of the flaws in its program, something NASCAR later changed, was that the drug Mayfield tested positive for was not identified.

Mayfield continued to deny any sort of drug abuse, refused to enter NASCAR's recovery program, and took his case to the courts. He would eventually lose his court battle but continued to deny any drug abuse and refused to enter into a recovery program. The drug was ultimately revealed to be methamphetamine, and in the ensuing years witnesses that included family members confirm that Mayfield had used the drug on many occasions. He never raced in NASCAR again and in 2011 was arrested for possession of methamphetamine during a raid at his home. Most of the charges were later reduced or dismissed, and in 2014, Mayfield pleaded guilty to misdemeanor counts of possession of stolen property and drug paraphernalia.

He is racing again, but his NASCAR days are far behind him.

While Mayfield's brush with NASCAR's substance abuse policy ended his career, another Cup driver who received a ban had a much happier ending and is still racing in NASCAR's top series today.

A.J. Allmendinger began his professional racing career in open wheel cars. He joined NASCAR in 2006, and after a brief stint with Richard Petty Motorsports, Allmendinger was hired by Team Penske, one of the top teams in NASCAR, in 2012, replacing Kurt Busch after that driver and the team parted ways. Allmendinger seemed well on his way to becoming a NASCAR star. It was all put on hold in July of 2012 when NASCAR announced that Allmendinger had failed a drug test and had been suspended. Allmendinger, and much of NASCAR, was shocked. Allmendinger later admitted taking Adderall, albeit accidentally, after he said he was told it was an "energy pill."

Allmendinger chose not to fight NASCAR. Instead, he enrolled in the Road to Recovery program. Penske released Allmendinger the following month, and a month later, in September, Allmendinger completed the Road to Recovery program and was reinstated by NASCAR. He would return to Penske in a part-time role with the team in the Xfinity Series. He capped off his recovery story with an emotional first NASCAR win for the team in June of 2013. Allmendinger would score his first Cup win in June of 2014 for his new team, and the one he currently races for, JTG Daugherty Racing.

NASCAR is much more involved in stock car racing than simply seeing cars racing in circles. There are rules to make the racing fair and safe, and those rules are constantly changing. Finding a driver to cheer for, and understanding all the layers that make up a NASCAR race, will help you enjoy all that much more. Soon, you will understand why a "loose" car needs more "wedge" and how a "short-pitting" car can make up spots on the track.

CHAPTER 3

VIRTUAL REALITY WITHOUT THE VIRTUAL

SO NOW THAT you have a pretty good idea of what's going on during a race, it's time to figure out the best way to actually watch a NASCAR race. Sure, you can sit on the couch, and later we will go over the best ways to see a race on TV, but for a first-time fan, there is nothing better than being there in person.

If you are near a large city that hosts a NASCAR race, close enough that you can sleep at home, then lucky you. Many fans, however, will travel some distance to see a race, and that means you will need someplace to stay. The sad fact is that many hotels near NASCAR tracks not only fill up long in advance, but many will charge inflated prices and require minimum stays during a race week. The trick is to book early, and don't try and stay in sight of the track.

If the particular track is near a large city, look to that large city for a place to stay. Dover International Speedway in Delaware, for example, is just over an hour south of Philadelphia. Getting a hotel room in Wilmington, which is in Delaware but closer to Philly, is much easier and cheaper than trying to find a room in Dover.

What should your distance limit be? Mine is sixty minutes, which usually takes about one hour to drive. I can oftentimes find places closer, but using the sixty-minute rule, hotel rooms at a reasonable price have never been an issue.

Advance planning, a little research, and a map will go a long way toward saving you money and allowing you to enjoy the experience that much more. Another source of finding a room is to ask the track when you buy the ticket; many times the track representative is a great source of all sorts of valuable information.

And about those tickets. The best place to get buy your ticket is directly from the racetrack. You can purchase them through NASCAR. com as well, but avoid ticket brokers and especially avoid the scalpers you will inevitably see on the side of the road near a track. Not only will you normally pay higher face value, but in many states, scalping is illegal. The more hands touch your ticket, the more profit must be made, so why not just go to the source?

Want to go all in? Then consider camping. Camping is one way to stay at the track, usually the infield, and it will give you an immersive experience that you won't soon forget. For many fans, camping is the only way to go. Some infields and camping areas around a track during a race weekend are legendary. Talladega Superspeedway, for example, has the "Big One on the Blvd," an event held in the infield on the Saturday night prior to the Cup race that features a Mardi Gras-style parade with drivers, along with concerts and other adult-themed activities including women wrestling for a pickle in a tub full of barbeque sauce (yes, seriously). Most tracks will have concerts and other activities during the race weekend.

In the infield, you'll see campers staying in RVs festooned in the colors of their favorite drivers and old-school buses painted to match a driver's race car, mixed in with traditional RVs and campers. There are banners, driver standees, kiddie pools, and even full-service outdoor bars.

Unscheduled driver appearances are also not uncommon, because drivers also stay in the infield, although in a special heavily secured area in luxurious RVs. The "owners' lot," as it's commonly referred to, is where drivers, and of course owners, stay. It allows drivers quick access to the garage and is their home away from home during the long NASCAR season. But drivers sometimes need to leave the track, and most often the only way to do that is through the infield. It isn't all that unusual for a driver to stop and interact with fans, if they have time. During the race weekend at Watkins Glen in August of 2016, then-reigning Cup champion Kyle Busch stopped on his way out of the infield at an RV that had his number on the side. He signed the number and thanked the fans for their support, much to fans' shocked delight.

When I first began working in NASCAR, I too camped in the infield. It was an experience I won't soon forget. To this day, when entering an infield early on a Sunday morning, lines of people, many nursing hangovers, can be seen shuffling to the showers in a scene that looks like it's straight out of a zombie movie. I always smile knowing how much fun they've had the night before. And that's the advantage of camping; besides being at the track, it's being able to party with thousands of other NASCAR fans.

The camping sites at tracks have fresh water, restrooms with showers, and all the other amenities found at a typical campground. Many tracks also offer camping packages. For the August 2016 race weekend at Michigan International Speedway, for example, the track offered a camping package for four that included a reserved camping site and admission to "The Deck," an area above turn 3 where fans can watch all the action on the track all weekend while being entertained by deejays, for $255.

Camping at a track is the ultimate tailgate, but if you decide for a more sedate experience, you can still tailgate. First, however, you have to get to the track, and that can be a challenge if you don't do it correctly.

There are few things I hate more than race-day traffic. It doesn't matter where the track is located— trying to put upwards of 100,000, sometimes more, people into a relatively small space can create traffic challenges for anyone. Some tracks (Daytona) are better than others (Kentucky), but if you try to enter a track at the wrong time, it won't matter how good the track is at traffic control, you will be sitting in a long line of cars waiting to park.

And a note about that parking. Every track, and I mean every track, has free parking in its vicinity. It may be a little farther from the track, but there will be free shuttles that run on a pretty consistent basis. Yes, there are businesses near a track that will offer parking; the closer they are, the more premium the price, and in fact most businesses near a track will close and use their parking lots for fans on race day, charging quite a bit for a parking spot and making a nice profit for the business. Save your money and use the free parking (did I mention that every track has free parking?).

Try to time your arrival for at least four hours, preferably more, prior to the green flag. In almost every case, this will allow you to miss much of the traffic. If getting to a track four hours early is impossible, then either plan on waiting in traffic or try to wait until about an hour prior to the green flag. The closer it gets to the green flag, the more people are already at the track, and normally, the less traffic there will be on the roads getting in.

If you do decide to get to the track early, don't worry, you won't be left wondering what to do.

There is always plenty to do before a NASCAR race. *Photo by Greg Engle*

All tracks have plenty of prerace activities. There is a midway with displays, driver appearances, concerts, and much more.

Marcus Smith, President and COO of Speedway Motorsports Incorporated, which owns nine tracks, eight of which hold NASCAR events, knows plenty about keeping fans happy.

"A NASCAR race is three things in one," he said. "You've got a race, you've got a state fair, and [you've got a] music festival all in one."

Something else you will want to do shortly after you get to the track is to secure something called a FanVision, which is a portable device that will give you content while you are at the racetrack. You'll get live audio, video, and stats. There are in-car cameras, replays, and the ability to listen to the drivers and teams during a race.

"NASCAR is the only sport that, hands down, gives you access to that kind of opportunity to listen in on the driver and the crew chief, the spotters, strategizing, it's incredible," Marcus Smith said. "An equivalent would be if you could listen in on the NFL, to the quarterback, the head coach, and the offensive coordinator decide what play they're going to run or not run. It's phenomenal."

When you get to your seats, you will notice large video display screens. These are usually in the infield near turns 4 and 1 (at Texas, there is the world's largest video screen known as "Big Hoss" along the backstretch, and Charlotte has a slightly smaller one along its backstretch).

Renting a FanVision shortly after getting to the track is a good idea. *Photo by Greg Engle*

The FanVision is a must-have during a race weekend.
Photo courtesy of Racing Electronics

These screens show in-race views that you see on TV, along with replays and tons of info that enhance the at-track experience. The FanVision shows these same views, and you can scan all the radio

Texas Motor Speedway has the world's largest video screen known as "Big Hoss," which can be used to show action during a race or other programming during a rain delay.
Photo by Jessica Rohlik

channels, as well. It is the same device you will see drivers, spotters, and crew members using.

Fans, crew members, and even drivers can be seen with a FanVision throughout a race weekend. *Action Sports Photography courtesy of Racing Electronics*

As of 2016, a FanVision rental was $59.95 at the track, but if you go online prior to the race, you can reserve a rental for $49.95 (as of 2016), and it's well worth it. I got a FanVision in 2011 and could not imagine covering a race without one.

There are other ways you can stay informed at a race. Though going to a NASCAR race for the first time can be a somewhat intimidating experience, if you can tag along with someone who has "been there, done that," they can help you navigate and enjoy the experience that much more.

Another great tool is your smartphone. Not only can you follow along on social media, but every major

Driver Kevin Harvick using a FanVision during a practice session while waiting in the garage getting adjustments. *Photo by Greg Engle*

NASCAR track has its own app with maps, a schedule, and a lot of other invaluable information.

Once you have your FanVision, you may still have a few more hours to spend before the green flag falls. Even if you have seats for the grandstands, most NASCAR tracks will allow you to get into the infield prior to the race, so head there. It's here that you can get up close to the pits, the garage, and see everything that goes on prior to the race. If you have a prerace garage pass, you can get even closer.

Many tracks will sell you a prerace garage pass; however, buying one is not the only way to get access. Many fans can get a garage pass through other means, such as sponsors. If you work for a company that is involved in NASCAR, ask if they can secure you a prerace garage pass. That friend who has been before? They may also have access to one. You can also sign up for a prerace garage tour offered by sponsors; however, you need to sign up for one of these as soon as you can once you get to the track, as these tend to fill up rather fast.

If you do score the coveted garage pass, there are two levels, cold and hot. That doesn't refer to the outside temperature, but rather what's actually going on in the garage. A garage that's "cold" is the most open. Cars are normally not moving. A "hot" garage, on the other hand, is an active one; cars are heading out to the track for practice and coming back in, and there is generally much more activity. It's during this time that fewer people in a garage area is best. If you want to know if a garage is "cold" or "hot," look for the red lights at each garage entrance;

If cars are going through the garage under engine power, the garage is "hot." *Photo by Greg Engle*

if they are flashing, the garage is "hot." A garage will usually go "hot" an hour prior to the race to allow crews to get cars lined up for the race.

Crew members, NASCAR officials, executives, and members of the NASCAR media corps all have credit card-sized credentials known as "hard cards" that allow access to just about anywhere they want to go.

On race morning, the garage will typically open a few hours prior to the green flag. Crew members will prep cars and roll through the inspection line one last time. During an inspection, you'll see a group of NASCAR officials using templates to check all manner of things on a car. They also use a Laser Inspection System to inspect the areas underneath the car.

NASCAR race cars undergo a great deal of scrutiny long before they even get to the racetrack. When a team completes the building of a car, it has to be certified by NASCAR. This is done at the NASCAR Research & Development Center in Concord, North Carolina, where the car and chassis are painstakingly inspected, and NASCAR takes meticulous records about each piece and part. The car is then outfitted with a Radio Frequency Identification (RFID) chip.

At the track, when that car rolls into inspection, the RFID is read by the laser, and the certification records for that car are pulled. NASCAR officials can then compare them with the car the team has presented.

There can be some small adjustments, but there are still tolerances that a car must fall into in order to pass, and be allowed to race. Teams will always try to push the limits allowed by NASCAR as they try to find every bit of speed. If they push the envelope too far, NASCAR will usually find it in inspection and force the crew to readjust the car as necessary. Cars also have to undergo an inspection prior to qualifying, and there are occasions when a car fails inspection and is delayed in making it to qualifying as the crew hurriedly makes the changes NASCAR has called for.

Starting in 2017, NASCAR makes crews who fail a part of inspection take their cars all the way back to the garage before starting the entire inspection process over again.

Cars will go through the inspection process several times during a race weekend.
Photo by Greg Engle

During a race morning, crews prep cars according to a checklist. On pit road, other crew members roll out and set up the large pit boxes and get the pit road area in front of their stall, the designated area known as the pit box, ready for the race.

Tip: If you are looking for a free souvenir on pit road on race morning, and there was a support race the day before, look along the pit wall on the trackside; there will usually be more than one race-used lug nut from the day before.

There are several things that happen prior to the race in addition to inspection. About two hours prior to the race, the drivers' meeting is held. Drivers and their respective crew chiefs gather with NASCAR executives and officials. This meeting is usually a crowded one, as VIPs, visiting celebrities, and a select groups of fans all attend. After the VIPs and celebs are introduced, drivers and crew chiefs are given the basic information for the day's race, and questions are taken.

Since relatively few fans ever get to attend a drivers' meeting, nearly all the tracks will pipe the meeting over the public address system and show video on the screens (which you of course can watch on your

Crews will often fail inspection. As an anonymous crew chief once said, "If they don't fail inspection the first time, the crew is not trying hard enough." NASCAR will make the crew effect repairs in the garage and go back through the inspection line again. *Photo by Greg Engle*

FanVision). Any driver or crew chief who misses the drivers' meeting will be penalized by NASCAR and forced to start at the rear of the field for the race. That's why fans wanting to see their favorite driver can usually be found near the drivers' meeting entrance and exit.

Garage Etiquette

You have a garage pass and are wandering around soaking in all the prerace activity. For a first-timer it can be a bit overwhelming. How

Crew members set up pit boxes along pit road on race morning at Martinsville Speedway. *Photo by Greg Engle*

do you make sure you can enjoy the experience? There are a few basic rules, unwritten and not, that fans need to know.

Don't ever actually walk into a garage stall. Those are reserved for crew. View the work going on in the garage across the aisle, the path kept clear for cars. Many times, there will be plenty of NASCAR officials to remind fans of this, but this is still a good rule to follow.

Speaking of crew members, give them the right of way. The garage area on race day can be a crowded place, and crew members have jobs to do. Many times you will see them carting around stacks of tires, fuel cans, and other equipment. Keep an eye out and give them way.

Garage stalls are working areas for crews, not for fans. *Photo by Greg Engle*

Hearing protection is good to have. Just because the engines for the race won't be fired until minutes before the race doesn't mean they will be silent until then. Many teams will be checking engines on race morning and the transmissions, as well. This means that not only will engines be idling, but they'll be revving, too. This can get rather loud.

Feel free to take plenty of pictures. If you want pictures of you and your friends beside your favorite driver's car, the best time to take it is while the car is waiting to go into inspection. I've even seen crew members offer to take pictures of a group while they wait. Whatever you do, though, keep your hands off the car, and never lean on it. The crew works hard to get their car race-ready. The same goes for pit road. There is a path you can follow behind pit row, but if a crew has their pit area set up, stay out. The same for the pit box. Once a team sweeps out and tapes off their pit box, they don't want people walking through it.

A NASCAR garage area can be very crowded on race day. *Photo by Greg Engle*

On the opposite side of the garage stalls, in the area where you should stand, the team's haulers are lined up. These specialty tractor-trailers carry two fully prepared race cars along with the support equipment needed for race weekend. A team hauler is where crews and the driver congregate when not on the track or working on the car. The haulers are lined up according to the driver's point standings, so the top driver's team will line up first, with the rest falling in behind.

A good time to get a picture beside a car is when it is lined up for inspection. *Photo by Greg Engle*

Behind the haulers near the pathway, many teams will set up stands with sponsor info and hero cards for fans to take. A hero card is an 8 × 10 card featuring a picture of the team's driver and the car. Many fans grab these and go in search of that driver for an autograph, which is easily the biggest thing fans hope to get during a visit to the race track.

Haulers are lined up in the garage according to point standings. *Photo by Greg Engle*

Driver's Autographs

An autograph. It's probably the most sought-after souvenir for a fan visiting a track, the proof that they interacted with their favorite driver and other drivers. A signature from a driver on your ticket, garage pass,

On the opposite side of the garage stalls is the area where you should stand. *Photo by Greg Engle*

t-shirt, or something else means that you met, or at least briefly interacted with, your favorite driver. Drivers are usually easy to spot in the garage. Most times they are surrounded by fans taking pictures and putting stuff in front of the driver to sign, and almost always the driver obliges.

NASCAR tries to be as fan-friendly as possible, and drivers know that once they reach the status of NASCAR driver there are certain obligations outside the car that come with it. They must meet sponsor obligations, NASCAR media obligations, and interact with fans. The more popular the driver, the more mobbed that driver will be in the garage area.

Drivers will almost always give fans an autograph. *Photo by Greg Engle*

Some tracks, like Daytona International Speedway, have viewing areas for the garages that face the infield. Fans can peer in on the inside of the garage, and drivers can interact with fans via a small opening at the bottom of the window. This enables fans who can't get into the garage the opportunity to get an autograph and allows drivers to sign things in a somewhat controlled environment. Other opportunities for driver-fan interaction can come on the midway outside the track prior to the race. Many tracks will sell ticket packages that include a driver meet and greet. Also, in the days leading up to the race, many drivers will appear at local events for their sponsor or charity.

So if you are in the garage, when is the best time to see a driver? And when is the best time to approach a driver? No driver is more popular in NASCAR right now than Dale Earnhardt Jr. He's been voted by the fans as the sport's most popular driver a record fourteen consecutive times—and counting—as of 2016.

"I think the best time to find me is during practice when I'm around the car," Earnhardt told me. "I'm going to come out of the hauler and get in the car about 10 minutes before practice. That is a great time."

"I'm going to get out of the car after practice and go to the hauler. That is another good time. I'm going to go sit in the hauler in between all that. I'm in that hauler all day. I don't really come outside the hauler much other than to get to the car. [But] those are good times.

"I walk to the bus [motor home] after practice and before practice; that is a good time. That is usually about thirty minutes before practice. Not every driver is the same. Some drivers get here sooner. Some guys maybe not as on time."

As Dale Jr. points out, just prior to and just after practice sessions are good times, as are the times when the drivers are hanging out behind the haulers. Just don't get too close to the back doors; crew members will be constantly going in and out of their workshop on wheels. Standing near the main aisle is usually your best bet.

Another area is the path drivers and crew chiefs use to get to the prerace drivers' meeting. Some tracks will use a red carpet and allow fans to line up. Drivers, unless they are late for the meeting, will usually stop and interact with fans. In fact, many drivers go to the meeting early for just that reason.

If you don't have a garage pass, one of the best places is the area between the back of the prerace stage and the exit of the garage area. Drivers will make their way from the garage to a designated area behind the stage to await driver introductions (something they cannot miss: if they do, they are penalized by NASCAR). The path to this backstage area is normally a good place to snag an autograph.

So when shouldn't you approach a driver? Never approach a driver when they are in the garage talking with their crew. If you are standing just outside the garage, the driver knows and will usually come out to sign when they can. One of the biggest no-nos, and something that could get you escorted from the garage, is when the driver is actually inside the car. Not even media members can do this. When a driver is in the car, it's a signal to all that they just want to be left alone. On rare occasions, a driver will motion for a fan to come over for a handshake or a picture, but unless a driver signals you over, stay away.

After driver introductions and prior to the drivers climbing into their cars, they will loiter around the car. Some don't mind being approached, but others do. In general, however, once driver introductions are over, it's time to head to your seat and time for the drivers to focus on the task at hand.

"There's always time when we're walking around casually, things like that," 2016 Daytona 500 champion Denny Hamlin told me. "Really, it's tough for us because we are so easy to access at all times with no

Drivers can be found on pit road prior to qualifying and the race. *Photo by Greg Engle*

locker room except for maybe our haulers, as that's our space. . . . It's not that we're trying to get away. We do have to have some time to ourselves to be professional. It's tough right before a race, honestly.

"I feel like each team should have a ten-foot square space around their car that is just designated for driver and their team," he added. "I think we could definitely work on that. I think a lot of tracks are working on that."

Whether you get an autograph or not, once prerace driver introductions are over, the infield area on the front stretch and pit road will be cleared. Time to head to your seats and get ready to experience a NASCAR race.

Before the stage used for driver introductions is pulled away, a prerace ceremony will take place. An invocation will be given, the National Anthem will be performed (usually with a flyover of military aircraft), and soon after the grand marshal of the race will give those "most famous words in motorsports"—"Drivers, start your engines."

For the next few hours the race will unfold, a winner will be declared, and the celebration in victory lane will take place. Many, but not all, tracks will allow fans in the stands to return to the infield after the race. You can get close to victory lane, grab a few race-used lug nuts along pit wall, and generally soak in the postrace experience. You should do something like this after the race, because after all, those thousands of people who came into the track now have to leave that track. Just like when you started the day, the wrong timing can lead to long waits in traffic. So, take your time and let the masses leave. Many fans will actually continue their prerace tailgating after the race, knowing that they aren't going anywhere fast.

How long should you wait? At least an hour is a good rule; however, some tracks are better than others at managing traffic. A little bit of research prior to your race excursion can prove invaluable when it's time to go home.

If you remain in the stands after a race and want to know the best way to avoid much of the postrace traffic, watch the roads in the infield.

After the race, drivers will leave soon after. They all use private planes at a nearby airport and are in a hurry to get home. If you look hard enough you can usually see a line of vehicles moving relatively fast out of the infield. Figuring out the road outside the track they are taking may lead you to what I like to call the "super-secret escape route." Almost every track has one, a road away from the main flow of traffic that allows drivers, teams, and NASCAR officials a quicker exit from the area. You won't find these routes listed anywhere, and no one from a track will acknowledge they are there, but if you look hard enough, you will see that indeed they are. Once you are in traffic, you will see a heavy law enforcement presence directing traffic. While they may prohibit you from going in a certain direction, if you can make it to the road being used as the "super-secret escape route," they usually won't stop you.

Speaking of a law enforcement presence, often a NASCAR race can involve alcohol. Have fun, but be responsible, and have a designated driver. On your way out, you will be passing law enforcement personnel, and I've seen many, many fans on the side of the road after a race undergoing a roadside sobriety test. While it's always a bad idea to drink and drive, it's an even worse idea when so many law enforcement officers are present.

Eventually you will make it home, no doubt exhausted, but hopefully with the pleasant memories of one of the greatest spectacles in all of sports, a NASCAR race.

CHAPTER 4

TV TIME

IF YOU CAN'T get to a NASCAR race in person, you will likely be watching it on TV. Or perhaps you will be attending a race in the future but will first see it on the smaller screen.

There was a time when seeing a race on TV was rare. Witnessing an entire race flag to flag was even rarer, and seeing it live was almost unheard of. In the early days, the only coverage of NASCAR races away from the track were via newspaper stories the following day.

In 1970, ISC debuted its own radio outlet, the Motor Racing Network (MRN). The network delivered live flag-to-flag coverage of top-tier NASCAR races and, by the end of the decade, as a package of taped highlights on a network sports program such as ABC's *Wide World of Sports*.

The 1979 Daytona 500 marked the first time a complete race was shown live from start to finish. A last lap crash combined with an on-track fight and a huge snowstorm in the Northeast led to huge ratings and a sudden interest by TV executives in broadcasting more races.

The growth of cable and more attention by the networks led to more NASCAR races on TV, and by 1985, all Cup races were shown either live or on tape delay. By 1989, nearly every Cup race was shown live start to finish. At the time, however, the races were shown on several networks, and fans sometimes had trouble figuring out just which network was showing the race that week. That ended in 2001, the first year of a multiyear contract that split the NASCAR coverage between Fox Sports and Turner Sports, which was part of NBC. This new arrangement cut out ESPN, the once-upstart cable network that had been broadcasting many of the races, although they did retain the rights to the Truck Series until 2002.

It was still somewhat confusing, as the new package called for certain races, such as those at Daytona, to be moved from network to network every year; Fox had the Daytona 500 in odd-numbered years, while NBC had the Daytona summer race; the following year they would swap.

Yes, it was a bit puzzling.

The arrangement continued until 2006, the first year in which a new contract started. Races were split among three networks. Fox carried the first thirteen races, TNT the next six, and ESPN the remainder.

Currently, there are two major networks that carry NASCAR—Fox and NBC. Unlike the years prior to 2007, every time a Cup car is on track, for practice or qualifying, it can usually be seen live on television.

Coverage during a race weekend (which can include Thursday and Friday) can be found on the various subchannels owned by the network. For Fox, that can be Fox Sports 1 (FS1) or, on rare occasions, FX or FS2. For NBC, it's the NBC Sports Network. Only the big races, like the Daytona 500 and the Brickyard 400, will be shown on the main network channel, but those occasions are rarer.

The Xfinity Series races follow the same template, with the races that take place up until the second Daytona race shown on Fox, and the rest on NBC. Fox currently carries all the Truck Series races in a season.

Most Cup races average around four hours. The Xfinity and Truck races, of course, take less time but still require a time investment of two or more hours.

The networks will always have a prerace show that will catch you up on the news of the week and important things to know leading up to the day's race. While the prerace show is normally on the same channel as the race itself, there are a few times a year when it's not. There's also a postrace show. Both the pre- and postrace shows are scheduled to last a half-hour, but for the bigger races, the Daytona 500 for example, the prerace show can be an hour or more.

The point is that watching a NASCAR race from start to finish will require an investment of time that can last anywhere from two to five hours. And if rain delays a race, especially prior to the end of the second stage (the point at which NASCAR can declare a race official), it could last much, much longer. I once arrived at Atlanta Motor Speedway at 8 in the morning and, thanks to rain off and on all day and the fact that the track had lights, left the next day at 4 a.m.

While that may seem like a significant amount of time to invest, as you have already learned, there is a great deal going on over the course of 400, 500 and once a year 600 miles. Track conditions will change, the race may transition from day to night, and someone who is leading or running near the front early on may not be there at the end. Consequently, your favorite driver may struggle early on but finish strong.

You are not left on your own during a TV broadcast. No matter which network is broadcasting the race, there is a great deal of preparation that all networks do prior to a race. The networks are a traveling show unto their own. The various trailers containing the equipment needed for a live broadcast make up the "TV compound"—a fenced area just outside the track that is the heart of the operation. The trailers consist of production facilities and backup generators. The mobile units usually arrive by Tuesday of race week. By Thursday, they are fully set up and ready to go.

That preparation prior to the race actually started months earlier. In fact, for those who work at Fox Sports, the network that carries the first half of the NASCAR season, that preparation starts the year before.

Barry Landis is a five-time Emmy Award-winner and the lead producer for all FOX NASCAR broadcasts. He is also a producer for the NFL on FOX. The 58th running of the Daytona 500 in February 2016 marked Landis's eleventh consecutive year producing the race (2006–16) and his fourteenth year overall working on the TV production side of the event. He was also producer and coordinating producer for NASCAR on TNT.

Landis, who joined FOX Sports in 1994, said their prep begins as the network completes the first half of the previous season. Currently, Fox ends its flag-to-flag coverage after the Cup race at Sonoma in June.

"Almost immediately we start working on things for the next season," Landis said. "Storylines aside, we have plenty of elements that can be worked on such as our graphics package, new technology research and development, as well as television compound logistics, to name a few.

"In reality, FOX NASCAR has been a year-round job since 2001. Many of our on-air talent are at the track during the second half of the season and NASCAR Race Hub on FS1 keeps everyone active within the sport. I would say that by the time December rolls around, we are all anxious to get down to Daytona. Our weekly talent and production calls about two months before Daytona, with the goal in a perfect world, to have our game plan finalized by the first week in January."

The preparation for each race during the season starts weeks prior.

"Once the season begins it is tough to get too far ahead of ourselves because each event is as important as the next," Landis said. "That said, we do look two to three weeks down the road so that we do not allow anything to slip through the cracks. It would be simple to say that we categorize races by tracks (superspeedways, short tracks, mile-and-and-a-half) but we are constantly looking to improve our coverage, so it would a disservice to the viewer if we were to say, 'We did this at Daytona so it will be the same at Talladega.' Just as NASCAR makes aero/mechanical adjustments, we have to adjust as well.

"The bigger factor in preparation during the season are the storylines," he added. "Who's hot? Who's struggling? You can't look too far down the road because the story may change. From a logistics and budget standpoint, we do map out what we think we will need for each race in the months leading up to Daytona, but we are prepared to adjust."

Those preparations intensify during the week leading up to the race; however, in some ways the preparations are fluid and continuous.

"Our statistical mining and analysis is constant," Landis said. "Our stats team from Racing Insights works tirelessly throughout the year to keep us all informed on trends. Our announcers (Larry McReynolds, in particular) keep a close eye on the race car changes throughout the year. We have an open forum all year long and keep each other informed.

"As far as setting up the compound is concerned, our technical operations department works with NASCAR Media Group well in advance to ensure an efficient setup," he added. "If there is going to be a change in compound location, we usually are made aware of the plan a year in advance."

Landis said the goal of all the preparation is simple:

"To give our viewers the best possible NASCAR experience on a weekly basis. It's not lip service. I don't think anyone on our team wants a pat on the back for how hard they worked to bring the event home to fans. They would rather be known as being the gold standard in sports television."

As for what goes on behind the scenes on race day, Landis gives some insight.

"Let's start with the beginning of the race weekend first," Landis said. "We televise every practice and qualifying session leading up to the race. I spoke earlier of being flexible and willing to adjust based on storylines. Well, practice and qualifying play a big role in doing so. We literally see every practice and qualifying lap, and it gives us a good idea of who looks promising for the race or who needs to get to work. Because we are at the track televising practices and qualifying, technically we are set up for the race."

"Our race day begins with an early meeting with NASCAR officials to answer/ask any questions that might be out there. Our production meeting immediately follows. All on-air talent and production people are present to review prerace show formats, as well as to discuss key elements we would like to touch on during the race. This meeting is the culmination of our preparation during the week. We leave that meeting with a plan or blueprint for the telecast, *but* the live event dictates our coverage. The meeting lets everyone know what we have and what we are thinking so we can react in the moment by putting our best foot forward.

"From a viewer's perspective, our prerace show will give you what to look for in the race along with a lot of entertainment. Our race coverage will tell the story as its unfolds with expert analysis and dynamic pictures and audio. The postrace will sum it all up with driver interviews and replays of key moments."

Landis is quick to point out that broadcasting a race is a shared endeavor.

"It's not a cop-out to say FOX NASCAR is a collaborative effort. No one person is in charge of our race day. Jake Jolivette and Bill Richards produce our prerace shows, and I produce the race action. Jeremy Green directs the cameras for the prerace and Artie Kempner directs the race. Our team of technical producers, as well as our technical operations group, handle facilities and manpower. The announcers (our experts) drive the content . . . the list goes on.

"Since not every viewer is watching for the same thing, it is important for us to work together to provide as many lenses to satisfy as many fans' needs as possible. To say that one person is in charge would be inaccurate."

The main person calling the race, the announcer, is not simply a "talking head" who is reading off a teleprompter. All the announcers for each network have extensive backgrounds in racing and have been around NASCAR for many years.

Mike Joy is a broadcast veteran with forty-six years of motorsports experience. He is currently the lead race announcer for Fox NASCAR. He works alongside analysts and former champions Darrell Waltrip and Jeff Gordon. Since 2001, the first year for Fox as a NASCAR broadcast partner, Joy has led the network's broadcast team.

Joy anchored CBS Sports' coverage of the Daytona 500 in 1998 and 2000. He was a pit reporter for fifteen years prior to that. Joy called the "Great American Race" for MRN Radio from 1977 to 1983. He has also hosted Formula One coverage for the Fox Sports Network and was a contributor to the 1992 Olympic Winter Games coverage. His coverage experience also includes CBS college football and NCAA championship events. He started his career as a PA announcer at Riverside Park Speedway in Agawam, Massachusetts, in 1970. He also anchored the first live NASCAR Cup telecasts on ESPN in 1981 and TNN in 1991.

"I grew up wanting to be the next Dan Gurney . . . he was the All-American racing hero," Joy said. "We're friends, and I'm still a huge fan. It didn't take long for me to find that I had a much better chance of making a living with a microphone than a steering wheel. As a sports car racer, I was a decent amateur, and I am very lucky to still get to scratch that itch by running some vintage races each year with the Historic Trans-Am group. I'd love to have one of Dan's former race cars, but, because he drove them, I can't afford them."

Despite an extensive knowledge of NASCAR, Joy still does a lot of preparation prior to the week's race.

"Lots of time on the Internet, plenty of time on the phone, some time face-to-face with drivers, owners, team members, and officials, a weekly conference call and one or more production meetings onsite," he said. "All the practice and qualifying shows we do are also good prep for the race."

Mike Joy spends a great deal of time getting ready to broadcast a race. *Photo courtesy of FOX Sports*

Thanks to his extensive knowledge and experience, Joy said he is allowed a great deal of input before and during a race.

"The booth drives our telecast, and while we go into each race with some storylines we think will be important, 90 percent of our coverage is reactive to what is happening on the track and predictive to what we think could happen next," he said. "Once the prerace show ends, there is *no* script."

Fans watching on TV hear Joy and the race analysts call the race, and fans may not realize that during the race the TV booth is a very busy place.

"I'm constantly having three conversations . . . one with Darrell, Jeff, and Larry, one with the viewer, and one through the talkback channel and IFB (interrupted feedback) in my left ear with Barry Landis, our producer, and occasionally with director Artie Kempner or coordinating producer Richie Zyontz," Joy said. "It's a three-ring circus, and that's all part of what makes this *so* challenging, and so much fun when you get it right. Unlike other sports that have *one* ball to follow, we have *forty* of them starting every race!"

The goal of all this challenging work is simple, Joy said:

"I'd like the viewer to remember more and interesting details than he can find online or in the paper the next morning," he said. "I hope he or she takes a few factoids about the race to throw out at the Monday morning water cooler conversation."

Broadcasters like Joy aren't alone in the booth. Every network uses a pair of color commentators, analysts, who are either former drivers or crew chiefs. Fox Sports uses former NASCAR champion Darrell Waltrip and newly retired champion Jeff Gordon. Combined, Waltrip and Gordon have more than 1,600 Cup starts, with 177 wins and seven Cup titles. They also use former crew chief Larry McReynolds (known as simply "Larry Mac") in various roles in the garage, or in pre- and postrace shows.

"They *are* the telecast," Joy said. "This is a sport of great complexity . . . strategy, mechanical innovation and advantage, hand-eye

FOX Sports uses former NASCAR champions Darrell Waltrip and newly retired champion
Jeff Gordon as analysts during a broadcast. *Photo courtesy of FOX Sports*

coordination, skill, stamina, even bravery (so said Hemingway). Darrell
and Jeff have seven NASCAR Premier Series championships in the
booth and, counting Larry Mac, some 200 race wins to draw from.
It's my role to take their knowledge and expertise and weave it into
a telecast that is informative *and* entertaining. I don't have to worry
about the entertaining part . . . they're gifted, and we *insist* on having
fun with our coverage . . . never making fun *of* the sport, but having
fun *with* it."

As for new fans of the sport, Joy has some advice for how to watch,
enjoy, and get the most out of a NASCAR race.

"Pick two or three drivers and try to follow their progress through
the race," he said. "Note when they are gaining on the leader or losing
time. Cheer for them when they are leading or making their way to the
front. Complain loudly to the TV when your drivers aren't being shown.
Wince at some of our historical and pop culture references . . . some are
actually interesting, some are purposely cringeworthy . . . keeps you on
your toes.

Jeff Gordon and Darrell Waltrip have more than 1,600 Cup starts, with 177 wins and seven Cup titles. *Photo courtesy of Jeff Gordon*

"And, take a break every once in a while and go to the refrigerator," he added. "As Warren Zevon said very late in life, 'Enjoy every sandwich.' Three and a half hours is a long time to perch in front of the TV."

Jeff Gordon had a chance to not only step out of the race car and into the booth, but in the middle of the 2016 season, shortly after his first stint in the Fox broadcast booth ended, Gordon was called out of retirement to serve as a sub (along with young driver Alex Bowman) for Dale Earnhardt Jr., who was suffering from concussion-like symptoms.

"It helped me being an analyst by getting out of the car, and being fresh out of the car from last year," Gordon told me prior to his final race as an Earnhardt sub, at Martinsville in October. "It gives you a much better perspective on what is happening. What the drivers are going through. Trying to put yourself in their position. Seeing things on the track that are happening and understanding what the aerodynamics are like. What the setups are like. What the grip is like.

"It was good to see the difference of how the cars drive this year versus how the cars drove last year. I give a lot of credit to these teams for what they have done with less downforce, a combination of Goodyear and the engineers and crew chiefs that build these race cars. I was pretty shocked that these cars really haven't lost anything."

Remember, watching a NASCAR race isn't something you can do in an hour or two. It will require an investment of your time on a Saturday night, or Sunday afternoon (when Cup races are normally run). Don't worry if you don't understand everything that is going on. Depend on the announcers, pit reporters, and the analysts who have a great deal of experience in a NASCAR race car. The announcers will keep you informed, educated, and, most important, entertained!

Here's an insider tip: if you live near a major NASCAR track, many have "watch parties" during the big races if they are not hosting the actual event. Watching a race with a group of fans is always a good time, and watching a broadcast at a track with fans only enhances the experience.

"Tune in and crank up the volume!" Barry Landis said. "If you have ever wondered what it's like to drive 200 mph in rush hour traffic, then our coverage will take you as close to that as possible. Don't be afraid that you don't have the knowledge to follow the action, because you will hear analysis that is informative and entertaining. Pick a car/driver (your favorite number, car manufacturer, hometown) and pull for them. By the end of the day, you will have more than one reason to tune in the following week."

CHAPTER 5
THE ONLINE EXPERIENCE

THERE WAS A time when NASCAR fans were happy to simply see racing on TV. When races began to be shown live flag to flag, no longer did fans have to rely on newspaper stories or a radio broadcast. Today, however, NASCAR fans are more connected to the sport than ever before. The NASCAR "lifestyle" extends beyond the weekends. Like the rest of the world, NASCAR is on a 24-hour, 7-day-a-week cycle, thanks to the Internet.

NASCAR launched its own website, NASCAR.com, in 1995. At first simply a place for the latest news, standings, and schedules, today NASCAR.com is the central hub for all things NASCAR. From news, to previews, lifestyle articles, and videos, the site also provides live timing and scoring during races, including in-car cameras and up-to-the-minute telemetry. The site's RaceView product delivers virtual race content, including live stats, lap-by-lap commentary, a live leaderboard, and audio, including the ability to listen to a team's scanner channel.

In 2001, NASCAR sold its digital rights to Turner Sports, giving them control of the content and management of the online content of the sport. At the time, the stigma that was associated with a major sport controlling its own online content still mirrored historical concerns about newspapers being a separate entity from the sports or topics that they covered. At first, this carried over to the emerging world of the Internet.

The deal between NASCAR and Turner Sports worked fine for a decade, but as more major sports took control of their own content and newspapers became less relevant, the landscape changed, and NASCAR.com changed with it.

NASCAR took back the rights to their digital content starting in 2013. Currently, Turner still handles advertising and sponsorship for the site, but today NASCAR has a fully staffed digital team. That team was part of a behind-the-scenes expansion of the NASCAR public relations staff that in 2011 became the Integrated Marketing Department.

The public relations arm of NASCAR grew as the sport did. For many years, people who had extensive experience primarily in newspaper reporting handled NASCAR PR. Several reporters gave up their typewriters and took over PR duties for the sport. Perhaps the last, and best known, of these was Jim Hunter.

Jim Hunter was a football and baseball player at the University of South Carolina in his youth. But he had an attraction to motorsports and soon was handling public relations for several top IndyCar drivers before taking on the role of public relations director at Darlington Raceway, the historic track the South Carolina native always called his home track. He later took on the same role at Talladega Superspeedway.

Hunter would go on to be the sports editor of the *Columbia* [SC] *Record* newspaper. He was also a columnist at the *Atlanta Journal-Constitution*. He joined the executive ranks of NASCAR in 1983 as vice president of administration. In 1993, Hunter eagerly accepted the offer of president at Darlington Raceway and was named a corporate vice president of the International Speedway Corporation. He returned to Daytona Beach in 2001 as vice president of corporate communications, leading the newly expanded public relations effort for NASCAR.

Jim Hunter, and NASCAR, knew the emerging world of the Internet would change the way NASCAR news was reported. Prior to his appointment as VP of communications, newspapers were king in the NASCAR media corps. Internet reporters were treated as almost second-class citizens. Rarely were press credentials issued to Internet reporters, and then they were highly restricted.

At the turn of the 21st century, website reporters most often had to do their job from afar or even in the grandstands, having purchased a ticket for the race.

That began to change under Hunter's stewardship.

NASCAR began to recognize that more of its news would be reported in media beyond newspapers. As the "face" of the NASCAR PR effort during this time, Hunter ensured that everyone in a media center was treated fairly and equally. Web reporters began getting better access and by 2005 were considered on a level playing field with their newspaper counterparts.

Jim Hunter passed away in 2010, a victim of lung cancer, at the age of seventy-one. He represented perhaps the last of the "old school." By the time of his passing, fewer newspapers had full-time reporters who traveled with the circuit every week. In 2016 only the *Charlotte Observer*, and the national newspaper *USA Today* had full-time staff dedicated to NASCAR coverage. tronc, Inc. (once known as Tribune Publishing), which owns several large newspapers, has a reporter at many races. However, that reporter is not dedicated solely to NASCAR.

After Jim Hunter, the NASCAR PR office—or, as it was called starting in 2011, the Integrated Marketing Communications (IMC) office—became less of a sports-focused PR firm and more of a marketer of the NASCAR brand. New staff swelled the ranks of IMC, with most bringing a great deal of experience from other professional sports. The formation of IMC signaled the completion of the transformation of NASCAR news from a newspaper cycle to one that catered to the digital world.

Today, more Internet reporters cover the sport than traditional newspaper reporters do.

Some of today's Internet reporters feed websites connected to newspapers, primarily *USA Today* and the Associated Press. Most others write for websites dedicated to the TV partners, NBC Sports and FOX Sports. There are also writers for ESPN, Yahoo Sports, and sites dedicated to auto racing, such as Motorsport.com. Other sites started as more fan-driven, which is to say they began as sites that at one time delivered content written by fans but now have full-time reporters; these sites include SBNation.com and Bleacher Report.

There's even a site staffed by former newspaper reporters who covered NASCAR for some of the largest newspapers in America but had no place to go when their jobs were eliminated. The site, Racintoday.com, has writers that at one time covered the sport for such papers as the *Atlanta Journal-Constitution* and the *Los Angeles Times*.

Both of the national syndicated radio networks that cover the sports, MRN (owned by ISC) and PRN (the Performance Racing Network, owned by SMI), have websites that not only provide news, but also live, free broadcast streams during a race. There are also hundreds of sites devoted to NASCAR run by fans.

None of these sites, however, are arguably as big, or can have as much of an impact on NASCAR, as a site started by a guy from New Jersey as a class project in the 1990s.

Jay Adamczyk is known simply as Jayski. His site, Jayski.com, has become the place for fans and those within the NASCAR industry for the latest NASCAR news and information.

An Air Force veteran, Adamczyk was working for the FAA when he decided to go back to school. As part of a class project, students were asked to build a live webpage on the new emerging medium of the Internet. Jay completed the assignment, building a tribute page to his favorite driver. NASCAR fans, lacking many online resources in the young days of the Internet, began to email him. Some asked questions about the sport, others shared news they had found. Intrigued, Jay began scouring the few resources on the Internet at the time, mainly newsgroups, searching for NASCAR news.

Soon those in the NASCAR industry also grew interested in the site, and Adamczyk began to get "insider" news about driver changes and other interesting tidbits. His small website began to grow, and in 1995, Jayski.com, or "Jayski's Silly Season Site," as it is also known, was born. The "Silly Season" refers to that time of the year, usually in the second half of the NASCAR season, when drivers and teams announce new contracts, or drivers announce they are going to a new team. Because of all the logistics involved in a driver moving to a new team, or the announcement of a new sponsor, no one can wait until the off-season. Teams and sponsors, wanting maximum impact, will want to keep an announcement a secret until they decide to reveal the news. These announcements usually occur during the "Silly Season." However, the more people involved with the logistics of a driver change or sponsor announcement, the better the chance that the information will be leaked.

It was these sorts of leaks that Jayski.com thrived on in the early days. Jay left his FAA job in 1999 to run the site full-time. From those early days to today, Jay has built the site to become a repository of NASCAR news and information. More than just verified and unverified rumors, driver and sponsorship news, the site now features paint schemes, TV ratings, history, updated radio frequencies, and just about every stat about NASCAR you can imagine. Although the site is now owned by EPSN, Jay still very much runs and controls it.

Like in the early days, Jay still remains relatively anonymous. In fact, for much of the early years, Jay's real identity was a mystery. It had to be. Some of the stories he broke on his website contained information that teams, sponsors, and NASCAR didn't want public. This didn't earn him many fans with NASCAR executives.

The argument can be made that the emergence of Jayski.com forced those in the industry to become much more guarded with their information. Today, however, those in the NASCAR industry are much more accepting of Jayski.com and understand how useful the site can be in disseminating NASCAR news and information.

Jay Adamczyk can now be seen on a rare occasion on TV, or at a race shop near his home in Charlotte, where he moved several years ago. One place he isn't seen, however, is at a race. Jay has rarely attended races, and that remains the same to this day.

For those of us who run NASCAR websites, Jayski.com is not only an important source of news and information, but a big source of page views, as well. One of the many things Jayski.com provides is an up-to-date page of links to NASCAR-related stories from around the Internet that Jay finds interesting.

If you want to use the Internet to know what's going on in NASCAR at the track and away, NASCAR.com and Jayski.com are the two places with which you need to start.

Today's fans want, and expect, their NASCAR news and information as close to real time as possible. If it's not being broadcast on TV, fans will search the Internet, and today's NASCAR media corps meets that need and feeds the Internet beast.

A deadline room at a track where members of the media work. The deadline room, like this one at Daytona International Speedway, is called that because at one time, newspaper reporters were on deadline. Today, however, the NASCAR news cycle is 24/7.
Photo by Greg Engle

Behind all this are NASCAR's social media and research and insights teams. In 2013, NASCAR unveiled its Fan & Media Engagement Center (FMEC) in Charlotte. The idea of Brian France, the center is responsible for monitoring digital and social media surrounding

In 2013, NASCAR unveiled its Fan & Media Engagement Center in Charlotte. *Photo by Scott Hunter for NASCAR*

the sport. The sanctioning body uses banks of monitors to see what is going on in real time in the digital world.

Operating seven days a week, during races the FMEC interacts with fans using the official NASCAR social media channels. Two staffers monitor trending topics and see what people are talking about. The FMEC also provides an analysis after the race of the social media engagement, citing such things as the number of social media posts, what spurred the posts, and how the volume of posts compared to past races.

Just as the Internet never takes a break, the FMEC constantly monitors social media even when there is no racing. The team gauges interest in NASCAR-related topics such as sponsor announcements, driver interactions, and other NASCAR-related happenings, providing analyses that can be helpful to those in the NASCAR industry. While not wanting in any way to control or restrict the coverage of the sport, NASCAR can expand their reach and elevate and amplify the conversation around the sport as well as be a part of the conversation while fans are watching the race.

Today, the Internet is king when it comes to NASCAR news, and social media not only delivers the type of instant news NASCAR fans crave, but connects them to the sport more than ever before.

In one recent example, driver Brad Keselowski took center stage during a red flag period at the 2012 Daytona 500, thanks to social media. The race had already been delayed a day due to rain and was under another long delay thanks to driver Juan Pablo Montoya hitting a jet dryer, causing it to explode in a fireball (thankfully, no one was injured).

Keselowski stood around with other drivers on the backstretch and began tweeting. He posted pictures and messages and interacted with fans. It wasn't long before his tweets began trending worldwide. It also got the attention of NASCAR. While they appreciated the exposure from a PR standpoint, the fact that drivers had cell phones with them in the race car during a race didn't go over well. The following week, NASCAR issued a rule that prevented drivers from having any sort of electronic device in the car with them, including cell phones.

Some of the more active drivers on Twitter, in addition to Keselowski, are Kyle Busch, Martin Truex Jr., Joey Logano, Kevin Harvick, and Dale Earnhardt Jr., who has millions of followers. Drivers' wives are also popular on Twitter. Samantha Busch, wife of Kyle, tweets a great deal, as does Delana Harvick, wife of Kevin. Crew chiefs can be found on Twitter along with various crew members.

"I think the biggest thing since social media, even the Internet, [is] the way we all consume the news is so much different than it used to be," 2003 NASCAR Cup champion Matt Kenseth told me.

"We used to have a ton of different beat writers show up at the racetrack, local newspapers, the normal NASCAR publications, all those people, and that's who you do interviews with, and that's how people would consume the news," said Kenseth, now forty-five, who entered the Cup series in 1998.

"I remember my grandpa and my dad, forever, everybody would have the *Winston Cup Scene* sitting at home, they couldn't wait to get it every week. That's how they'd read about the races and everything that goes on. Now everything is just instantaneous. The NASCAR writers are all on the Internet. You can get on a website and get an article right away, a tweet to a link to an article right away. The reporters tweet about what happens.

"The drivers, the crew chiefs, everybody seems to tweet or post stuff in different places. . . . Nobody wants to wait for the news. So certainly, the way it's consumed is different.

"The way we interact with fans is probably a little bit different because of social media. You know, sometimes it's good, sometimes it's bad. They can send you their thoughts and messages, pretty much right away. Sometimes we use that, sometimes we don't, but that's certainly changed things."

There are also social media accounts for teams, administered by public relations representatives, who give fans a behind-the-scenes look at the NASCAR world.

None of these PR reps is more famous perhaps than Bryan Cook, a.k.a. "Boris" from Joe Gibbs Racing, who serves as the team's director of social media. The lifelong race fan raced street stocks in high school at Hialeah Speedway, near a suburb of his hometown, Miami, the same track where legendary NASCAR driver Bobby Allison cut his teeth.

With an interest in visual arts and design, Bryan created a website for a young driver, Hank Parker Jr. The driver liked what he saw, soon the two became friends, and Parker became his mentor into the world of NASCAR.

Bryan moved to Charlotte after graduating high school to get close to NASCAR. There, he attended the University of North Carolina at Charlotte, earning his degree in art. While working as a freelance graphic designer, he got the call that started his career in motorsports.

"For three years," Bryan said. "I was a freelancer designer for Chip Ganassi Racing and designed the paint schemes on over twenty cars that appeared on-track, for drivers like Juan Pablo Montoya, Casey Mears, and Reed Sorenson. That was the thrill of a lifetime, seeing the cars on track (and in diecast form!).”

By 2009, social media was starting to emerge as an important tool for marketing, and marketers in NASCAR began to take notice.

“A person I went to church with told me that Joe Gibbs Racing was looking for someone to create content and make their web presence more compelling,” Bryan said. “It was a perfect fit, and I started traveling to every race almost immediately to cover the events from an [insider’s] perspective, specifically on Twitter during those days. I believe I was the first hire in the industry that was solely tasked with focusing on social media. Most others who dealt with it wore multiple hats like PR reps. That speaks to the foresight of Joe Gibbs, Team President Dave Alpern, and VP of Communications Chris Helein.”

The 2016 season was the seventh for the thirty-one-year-old, and indeed many in the industry credit him with showing how social media should be done. In fact, even NASCAR has called on him to help on occasion as a guest on their Snapchat account. To those in the industry and fans, however, he isn’t Bryan, but “Boris,” a nickname given to him by driver Joey Logano when Bryan first started working in the sport. With his long curly hairstyle at the time, Bryan resembled driver Boris Said, who is famous not only for his road course expertise, but his Afro, as well.

Bryan, a.k.a. Boris, has given fans insight and behind-the-scenes glimpses into the sport that no had done before.

“I still pinch myself to this day, in my seventh season, that I have the job I do,” he said. “To me, it’s the perfect merging of my passions for creativity, design, cars, and racing, and I feel blessed to have the opportunity I do. I always approach my content strategies from that standpoint.

“As a kid, I would have done anything to have the insight social media provides to fans these days,” he added. “To me, it’s important to keep in mind that social media marketing is novel because it’s on the fans' terms. We aren't 'shouting' content at them anymore, expecting them to watch or consume what we put out, on our terms. It's about connecting and engaging, with the goal of building new and stronger advocates of our team, drivers, and partners. My team and I have to be in tune with our fan base, posting content they will find appealing, at a level of quality and in a manner that Joe Gibbs would approve of and be proud of."

As Bryan explains, social media isn't just about connecting with fans. These days, social media has grown to become an added value to sponsors in NASCAR.

Many of the PR reps report to the sponsors on a regular basis. These reports include the value of the exposure the sponsor gets, not only on TV during a race, but on an ongoing basis away from the racetrack. Social media, something that was only a novelty just a few years ago, is now a huge part of the marketing plan for a team and a sponsor.

"It's an interesting time to be in this space," Bryan said. "I've always thought, there have to be dozens of people doing what I'm doing in other industries from politics, to entertainment, to sports, that have as many stories as I do. I have to believe that NASCAR is a unique animal, though. In our sport, when a corporation partners with a race team, they become the team. It isn't akin to purchasing space on a billboard or in a newspaper, it's a true partnership, even an emotional investment in many cases. I've had the opportunity to meet some of the most esteemed business leaders in the world because of the nature of NASCAR and because of the importance of social media."

As social media has grown in importance to the NASCAR industry, teams have needed to grow and adapt to the emerging medium. Most NASCAR industry insiders will credit the young man known as "Boris" with setting the social media standard that teams now aspire to. Drivers still post much of their own content, but today many have the assistance of their PR reps, ad agencies, or staff from the team whose specific job is social media. The trend started in large part by Bryan Cook. Today, social media gives fans greater exposure to their favorite

Bryan Cook, a.k.a. "Boris," has unprecedented access, giving fans a unique insider's view of the sport. *Photo by Jessica Rohlik*

drivers than ever before. Fans also get an insider's look that can only be provided by social media gurus like Boris.

"With the importance of social content (and by that I mean high-quality social content) increasing, my role with the drivers has become more integrated out of necessity," Bryan said. "I have access and relationships . . . that are not readily available to others. I fly with the team, have access to haulers, and meetings, and the shop, and that's where you find the good content people want. It's becoming increasingly important for drivers to have first-person content, so my involvement with drivers is becoming more and more about consultation and assisting them in making their own content happen."

Then, of course, there's also the NASCAR-operated platforms to consider.

In the middle of 2016, according to NASCAR, there were more than 114 million total engagements on NASCAR social platforms. As of June of 2016, the NASCAR Facebook and Twitter accounts had generated more than 2.3 *billion* impressions. Across all NASCAR digital platforms, NASCAR Digital Media had registered 507 million page views and 155 million on and off platform video views.

While the Internet has changed the way NASCAR news is disseminated and viewed, it has also changed the way NASCAR fans watch races. NASCAR.com provides a live leaderboard, along with a "lap-by-lap" commentary. The "RaceView" feature is a premium paid service that has customizable 3-D virtual views, radio broadcasts, telemetry and data, as well as live leaderboards. In 2016, the price for a full season of RaceView was $29.95, or $7.95 per month. The major networks also have companion sites that stream much of the same information along with different camera views. For those sites, however, you have to be a cable subscriber.

Is it worth paying for these premium features?

There are plenty of NASCAR fans who subscribe to "RaceView," and while Americans are becoming more and more disconnected from cable, many of us remain cable subscribers.

For many years, those in the media were given access to earlier versions of "RaceView" for free. During that time, I used it during every race. However, a few years ago that free access was taken away. Today, I find that using the free features on NASCAR.com, primarily the live leaderboard and the "lap-by-lap" commentary, are useful enough to save the money it would cost to subscribe to "RaceView." Many fans, though, couldn't imagine watching a race without "RaceView," so it's really up to individual preference.

NASCAR.com will on occasion have a free weekend of "RaceView" that can allow you to see if paying for a subscription will be worth it. These free weekends normally occur during the big races, the Daytona 500, The Coca-Cola 600, and the Brickyard 400, for example.

Staying connected when you are not connected

You already know that you can "see" much of what goes on during a NASCAR race without actually watching it on TV. Using online resources such as Twitter and NASCAR.com, you can get live in-race updates. After the race, NASCAR posts a video wrap-up and highlights on its YouTube channel, so seeing what happened after it's over isn't really an issue. The best part is that all this is available on today's smartphones. The NASCAR RaceView mobile app can be downloaded to your phone and offers the same features as the online computer desktop version for RaceView subscribers.

Both MRN and PRN also have apps. Neither of the radio broadcasters charges for their service, so listening to a race whether at the track, at home, or away has never been easier.

In fact, many fans use the RaceView feature as a companion while at the track. It may not provide the same sort of live camera views that the separate FanVision does, but if you are already paying to subscribe to RaceView, you can save the money on renting a FanVision at the track.

On the rare occasion when I'm not at the track, and on the even rarer occasion when I am not watching the race somewhere, you will see me holding my smartphone with my ear buds firmly in place. No, I am not listening to my favorite music, I am on NASCAR.com mobile, watching the live stats and listening to the lap-by-lap commentary via

My race day press box setup. *Photo by Greg Engle*

either MRN or PRN. Once the race is over, I'm watching the wrap-up and highlights on YouTube.

Accessing NASCAR today has never been easier. With all the races available on the Internet and Android and Apple apps, fans can keep up even when they are far away from the track and a TV.

Using the Internet and mobile apps can enhance a race experience both at the track and while watching a NASCAR race at home. Although some fans will pay the premium prices for the premium features, there are plenty of ways to get real-time information for free. Bookmark NASCAR.com and Jayski.com; from there you will discover that the digital world has a lot to offer when it comes to NASCAR.

CHAPTER 6
THE ULTIMATE FAN EXPERIENCE

IN ADDITION TO the fact that today's NASCAR media has never been more connected, the fans are also more connected today than ever before, and therefore able to enhance their own at-track or home race viewing by utilizing even just part of the technology that's offered today.

Many fans at home today will listen to a live radio broadcast online while watching the race on TV with the TV sound muted. Their tablet, desktop computer, or smartphone is on RaceView or NASCAR.com. On race day, Twitter is alive with drivers, teams, track PR reps, NASCAR executives, reporters, and fans.

I, like many reporters, will live tweet prior to, and during, the race, allowing fans a look on what goes on behind the TV screens. Drivers will tweet prior to the race, and many will have their PR reps tweet about their progress during the race (the tweet will usually end with something like –PR, so you know it isn't the actual driver). Track PR reps will usually inform fans at the track about happenings at the track that day, weather warnings, and other important info. NASCAR executives, meanwhile, have large followings, as they will usually inform fans and industry members about delays, the reasoning behind a certain ruling, or other information pertinent to the racing.

On race day, a few of the more popular reporters will stage a "Tweet-Up" hours before the race. This is a meeting at a designated area on track property where Twitter followers can meet in person. At a "Tweet-Up," fans can meet the reporters they follow on Twitter, and many times a driver, track president, or even a celebrity who might be in town for the race, or participating in prerace ceremonies. If you are at a race, and on Twitter, a "Tweet-Up" can be one of the many highlights of your race day experience.

If you are at home, reporters, NASCAR.com, and PR reps will live Tweet during a race, keeping you advised of up-to-the-minute action. That's because television broadcasters have to "pay the bills," so during the commercials, Twitter is the place to know what's going on. NASCAR is a live sport, and there are times when something will happen during a commercial. With Twitter, you'll always know what's going on.

Twitter is also a good place to find out what is happening with drivers deeper in the field.

The truth is, TV can only show so many stories, and those have to do with what is going on at the front of the field. Unless it's a star driver, say Dale Earnhardt Jr., what's happening outside the top 15 will usually escape the attention of the TV broadcast. If you are a fan of a lesser-known driver (and plenty of fans are), Twitter will keep you up-to-date.

Also remember this: TV isn't really "live." There is a delay that can range from 8-12 seconds. With Twitter and other social media platforms, fans can keep updated, even before it appears on TV. Social media can get you even closer than TV, and give a bigger picture of what's going on during the race.

Road Trip!

The best way to get closer to the sport, though, is to plan the ultimate road trip. If you are a fan of NASCAR, new or old, you need to make a pilgrimage to the heart of NASCAR country, Charlotte, North Carolina, at least once. Nearly all the NASCAR teams in existence today are headquartered in the Charlotte area. Most welcome fans to stop by for a visit, and during the week leading up to one of the two races at Charlotte Motor Speedway each season, many of the bigger teams will hold a Fan Fest with driver appearances, show cars, tours, and more.

For fans looking for deals on NASCAR-themed merchandise, the second round of Fan Fest every year, usually in October, is a great time to find deals on team gear, as this is when many of the teams hold clearance sales.

For new NASCAR fans, a trip to Charlotte needs to start at the NASCAR Hall of Fame.

For new NASCAR fans, a trip to Charlotte needs to start at the NASCAR Hall of Fame.
Photo courtesy of nascarhall.com

The process of constructing the NASCAR Hall of Fame began in 2005. Several cities were nominated, including Atlanta, Daytona Beach, and Charlotte. There was a short period of campaigning by each city, but few had any doubt that Charlotte would be the winner. Ground was broken in the spring of 2006, and the 150,000 square foot Hall in downtown Charlotte opened to the public in May of 2010.

The first class of inductees was enshrined the same month. Every year in May since, five more inductees are chosen from a field of nominees and inducted into the Hall in January.

The hall is much more than just a display for inductees. The multilevel museum contains high-tech interactive exhibits, a theater, and displays showing the history of the sport, as well as what goes on today.

The NASCAR Hall of Fame has plenty of interactive exhibits.
Photo courtesy of nascarhall.com

Located directly across from NASCAR headquarters in the heart of downtown, the Hall of Fame is an ideal place to learn about NASCAR's heritage and to gain an appreciation of what goes on in the NASCAR industry.

While the NASCAR Hall of Fame is a great first place to visit, there are so many other NASCAR-related places to see. You will need at least a few days to take it all in.

All the major NASCAR teams have shops that allow access to the public. They have common areas that let fans peer into the work area of the shop,

A visit to the NASCAR Hall of Fame is a great way to learn about NASCAR.
Photo courtesy of nascarhall.com

and most are open during regular business hours. Just a few years ago, most shops had only the working area and the business offices. Back then, while fans weren't necessarily prohibited from stopping by, they weren't exactly always welcomed. As NASCAR grew in popularity, however, that all changed. As new shops were built, areas for fans were included.

Penske Racing has a large open area with a fan walk overlooking the entire race shop.
Nigel Kinrade Photography for Penske Racing

Today, Joe Gibbs Racing in Huntersville, for example—not far from Charlotte Motor Speedway—will allow fans to peer in on an actual working shop from a second-floor display window. Trophies from race wins and championships are on public display, along with show cars and a gift shop.

The lobby of the Joe Gibbs Racing shop, like most race shops, is open to the public.
Photo by Greg Engle

The JGR shop is just one example, though. All the shops for major teams are set up this way, and a visit to the one for your favorite driver, or several, is well worth it.

There are numerous companies that charge a modest fee to take fans on a shop tour, but should you spend the money? Armed with a map, addresses, and some time, you can accomplish the same thing, without a fee.

There is no bad time to visit Charlotte if you are a NASCAR fan. However, there are some times of the year that are better than others are. During the month of May, Charlotte Motor Speedway holds two big Cup races. The first is the annual All-Star Race. The All-Star Race is a non-points event limited to a select group of drivers who have won a race

Most race shop lobbies are staffed and are glad to answer questions and hand out free Hero Cards. *Photo by Greg Engle*

the prior season or the current one, in addition to past All-Star winners and drivers who win their way in from the qualifying race or fan vote.

A week after the All-Star Race, the longest race of the year, currently called the Coca-Cola 600, is held on Memorial Day weekend. The week between these two races is a busy time in the area, and a busy time at the shops. Many hold fan fest-type activities, and Charlotte stages its annual Speed Street festival held by the 600 Festival Association. There is music from nationally known acts, displays, a hauler parade, and all sorts of activities on the downtown streets and surrounding areas. Many of the race shops during this time will also have driver appearances and sponsor activities.

In the fall (usually October), Charlotte also holds a second Cup race. The week leading up to that race is a good time to visit the shops for not only fan activities, but to find bargains on fan merchandise, as well. Many teams hold clearance sales in the fall, and this is a great time to visit and save a little cash.

There are other places to see, and things to do, that are NASCAR-related in the Charlotte area.

In addition to the Hall of Fame and real working NASCAR shops, there is the "Dale Trail." The "Dale Trail" is a self-guided tour around Kannapolis, a small town just north of Charlotte, and the hometown of the legendary Dale Earnhardt Sr. The tour takes you through the neighborhood where he grew up, along the streets he drove as a teenager, and to the larger-than-life bronze statute erected in his honor.

The "Dale Trail" ends at Dale Earnhardt Incorporated, the shop that was once a working one and is now a large showroom dedicated to the memory of one of NASCAR's icons. The huge DEI facility is now known as the "Garage Mahal," a shrine to the memory of Earnhardt Sr. containing trophies, show cars, and memorabilia from his career. For a new fan, the impact of Dale Earnhardt Sr. on NASCAR can be better understood through the history on display, while more seasoned fans will gain a better appreciation of the life of one of NASCAR's greatest drivers.

None of the race shops charge any sort of admission, and the self-guided "Dale Trail" only requires an investment of time. The NASCAR Hall of Fame does charge admission; however, most fans agree that the price is well worth it. Another museum that charges, but is well worth it, is the Richard Childress Racing Museum about an hour north of Charlotte in Welcome, North Carolina. Richard Childress was Dale Earnhardt's longtime team owner. Together, the two won six of Earnhardt's seven championships.

Richard Childress Racing remains one of NASCAR's major teams. The campus has working race shops with free access, but the museum charges a small admission fee. The museum is actually the first RCR shop built in 1986 that was turned into a museum in 1991. Not only do visitors see historic race cars, but a walk along the guided path through the museum is actually a walk through what was once an actual working shop. It's here that fans can learn what goes on in modern race shops. Each area mirrors what you will see in a working shop with explanations of just what is happening.

For a new fan, a visit to the RCR Museum is a great learning experience; it will help you understand just what goes into the making of a

NASCAR race car. From the fabrication of the body, to the assembly of the engine, to the painting of the body, to the haulers that take it to the track, it can all be seen under one roof.

For fans who know about the sport, the museum's curator is Danny "Chocolate" Myers, a longtime crew member for RCR and Earnhardt during the glory days. While his days crewing cars on pit road are behind him, Chocolate is still a popular figure in NASCAR. He spends a great deal of time at the museum and loves nothing more than meeting fans, posing for pictures, and telling NASCAR stories.

One exit north on U.S. Highway 52 is the Richard Childress Winery, owned by the famous team owner. Opened in 2005, the 35,000-square-foot facility is a working winery with a vineyard, tasting room, store, and bistro.

Visiting Charlotte during the week leading up to a race means that not only will you be visiting the Hall of Fame and race shops, but you will need tickets to the actual race.

So just what is the best way to get a NASCAR ticket? The answer is simple—go directly to the track you want to attend. Whether you use the Internet, the phone prior to your arrival, or you simply walk up to the ticket window when you get in town, buying tickets directly from the track is the best and safest way to get a seat. It is usually the least expensive way, as well. Searching for tickets on the Internet will many times lead you to a ticket broker who tacks on fees and charges much more than face value.

There have also been instances of counterfeit tickets being sold, something you won't know until you reach the gate on race day. Calling a track will not only ensure you are getting a genuine ticket, but a track representative can help with information about finding a hotel, parking, and traffic on race day.

Tickets that come in the mail often contain track maps, tip sheets, and a complete schedule of the weekend's events. Many tracks also offer combo specials that can save you money and add on special events such as prerace driver appearances.

Many tracks will have ticket specials you can only get directly from the track.
Photo courtesy of Texas Motor Speedway

These days, sadly, very few NASCAR races are sold out. The earlier you buy your ticket, however, the better chance you have of getting preferred seats, and the less chance you have of being shut out. Again, NASCAR races are usually not sold out, but if you hear they are, call the track. There could be last-minute cancellations, or tickets available as part of a combo package.

On the days leading up to the race, you'll see many people near the track holding signs saying they want to buy tickets, or have tickets to sell. Most often, these folks are simply looking to make a quick buck while NASCAR is in town. Scalping is illegal in many states, and the disappointment of finding out your ticket is a fake when you get to the gate isn't worth the gamble.

Combining TV coverage with the Internet will help immerse you in the world of NASCAR. Making a trip to the heart of NASCAR country, Charlotte, North Carolina, can not only show you how much work goes on behind the scenes, and how much work goes into getting a race car on the track, but can help new fans and seasoned ones gain

Imagine your disappointment when the person at the gate tells you your ticket is counterfeit, especially if that person is Eddie Gossage, president of Texas Motor Speedway.
Photo courtesy of Texas Motor Speedway

a better appreciation of history and traditions that make NASCAR a lifestyle for many.

Merging a trip to Charlotte with a race doesn't have to break your bank. Buy your tickets early, directly from the track, and consider visiting shops on your own.

There's a lot more to NASCAR than just a weekly broadcast. With a little planning, and armed with some basic information, you can get closer to NASCAR than ever before.

POSTSCRIPT
LIVING THE NASCAR LIFESTYLE

HOPEFULLY, YOU ARE now armed with enough information to understand the sport loved by millions. NASCAR is more than just a sport, however; it's a lifestyle. It's not perfect, but no competitive sport is; as those of us in the industry like to say, "Sometimes you'll have that in the world of big-time stock car auto racing."

Now put down the book (but keep it handy for reference). Turn on the TV, fire up the computer, or head to the track. Because on any given weekend during the year, there is a NASCAR race somewhere, and now you can be a part of, and enjoy, the world of big-time stock car auto racing.

NASCAR MEDIA WEBSITES/ TWITTER HANDLES

NASCAR.com
(@NASCAR)
Jayski.com
(@Jayski)

Foxsports.com/nascar
(@NASCARONFOX)
Main reporter: Tom Jensen (@tomjensen100)

NBCsports.com/nascar
(@NASCARonNBC)
Reporters: Nate Ryan (@nateryan)
Dustin Long (@dustinlong)

ESPN.com/racing/nascar
(@espnmotorsports)
Reporters: Bob Pockrass (@bobpockrass)
Marty Smith (@MartySmithESPN))

Usatoday.com/sports/nascar
(@USATODAYsports)
Reporters: Jeff Gluck (@jeff_gluck)
Mike Hembree (@mikehembree)

Sportingnews.com/nascar
(@sn_nascar)
Reporter: Rea White (@reawhite)

Charlotteobserver.com/sports/nascar-auto-racing
(@theobserver)
Reporter: David Scott (@davidscott14)

Motorsport.com/nascar
(@Motorsport)
Reporters: Jim Utter (@jim_utter)
Lee Spencer (@CandiceSpencer))

Sports.yahoo.com/nascar
(@YahooNASCAR)
Reporter: Jay Busbee (@jaybusbee)

MRN.com
(@MRNRadio)
Reporter (web): Pete Pistone (@ppistone)

GoPRN.com
(@PRNLive)

CupScene.com
(@cupscene)
Reporter: Greg Engle @GS_Engle

NASCAR tracks (2017)

Track/ Series	Length (miles)	Shape	Type	Address/Phone/ Web
Atlanta Motor Speedway Cup	1.540	Quad-oval	Intermediate	1500 Tara Pl, Hampton, GA 30228, www. atlantamotor-speedway.com/, (770) 946-4211, Twitter: @amsupdates
Auto Club Speedway Cup Xfinity	2.000	D-shaped Oval	Superspeed-way	9300 Cherry Ave, Fontana, CA 92335, https:// www.autoclub-speedway.com, (909) 429-5000, Twitter: @ACSupdates
Bristol Motor Speedway Cup Xfinity Truck	0.533	Oval	Short	151 Speedway Blvd, Bristol, TN 37620, www. bristolmotor-speedway.com, (423) 989-6933, Twitter: @BMSupdates

Canadian Tire Motor-sport Park Truck	2.459	Road Course	Road Course	3233 Concession Road 10, Bow-manville, ON L1C 3K6, Can-ada, canadiantire-motorsportpark. com, +1 905-983-9141, Twitter: @ctmpofficia
Charlotte Motor Speedway Cup Xfinity Truck	1.500	Quad-Oval	Intermediate	5555 Concord Pkwy S, Con-cord, NC 28027, www.charlotte-motorspeedway. com, (704) 455-3200, Twitter: @CLTMotorSpdwy
Chicago-land Speed-way Cup Xfinity Truck	1.500	D-Shaped Oval	Intermediate	500 Speedway Blvd, Joliet, IL 60433, www.chi-cagolandspeed-way.com, (815) 722-5500, Twitter: @ChicagolndSpdwy

Darlington Raceway Cup Xfinity	1.366	Oval	Intermediate	1301 Harry Byrd Hwy, Darlington, SC 29532, www.darlingtonraceway.com, (843) 395-8900, Twitter: @TooTough-ToTame
Daytona International Speedway Cup Xfinity Truck	2.500	Tri-Oval	Superspeed-way	1801 W International Speedway Blvd, Daytona Beach, FL 32114, www.daytonainternationalspeedway.com, (800) 748-7467, Twitter: @DISUpdates
Dover International Speedway Cup Xfinity Truck	1.000	Oval	Short	1131 N Dupont Hwy, Dover, DE 19901, www.doverspeedway.com, (855) 598-4722, Twitter: @MonsterMile
Eldora Speedway Truck	.500	Oval	Dirt	13929 Ohio 118, New Weston, OH 45348, www.eldoraspeedway.com, (937) 338-3815, Twitter: @EldoraSpeedway

Gateway Motorsports Park Truck	1.25	Oval	Intermediate	700 Raceway Blvd, Madison, IL 62060, www. gatewaymsp. com/, (618) 215-8888, Twitter: @GatewayMSP
Homestead-Miami Speedway Cup Xfinity Truck	1.500	Oval	Intermediate	700 Raceway Blvd, Madison, IL 62060, www. gatewaymsp. com/, (618) 215-8888, Twitter: @GatewayMSP
Indianapolis Motor Speedway Cup Xfinity	2.500	Oval	Superspeedway	4790 W 16th St, Speedway, IN 46222, www. indianapolis-motorspeedway. com, (317) 492-8500, Twitter: @IMS
Iowa Speedway Xfinity Truck	0.875	Oval	Short	3333 Rusty Wallace Dr, Newton, IA 50208, www. iowaspeedway. com, (866) 787-8946, Twitter: @iowaspeedway

Kansas Speedway Cup Xfinity Truck	1.500	Tri-Oval	Intermediate	400 Speedway Blvd, Kansas City, KS 66111, www.kansasspeedway.com, (855) 573-1351, Twitter: @kansasspeedway
Kentucky Speedway Cup Xfinity Truck	1.500	D-shaped oval	Intermediate	1 Kentucky Speedway Blvd, Sparta, KY 41086, www.kentuckyspeedway.com, (859) 578-2300, Twitter: @KySpeedway
Las Vegas Motor Speedway Cup Xfinity Truck	1.500	D-Shaped Oval	Intermediate	7000 N Las Vegas Blvd, Las Vegas, NV 89115, www.lvms.com, (702) 644-4444, Twitter: @LVMotorSpeedway
Martinsville Speedway Cup Truck	0.526	Oval	Short	340 Speedway Rd, Ridgeway, VA 24148, www.martinsvillespeedway.com, (276) 956-7200, Twitter: @MartinsvilleSwy

Michigan International Speedway Cup Xfinity Truck	2.000	D-shaped Oval	Superspeedway	12626 US-12, Brooklyn, MI 49230, www.mispeedway.com, (517) 592-6666, Twitter: @MISpeedway
Mid-Ohio Sports Car Course Xfinity	2.4	Road Course	Road Course	7721 Steam Corners Rd, Lexington, OH 44904, www.midohio.com, (419) 884-4000, Twitter: @Mid_Ohio
New Hampshire Motor Speedway Cup Xfinity Truck	1.058	Oval	Intermediate	1122 New Hampshire 106, Loudon, NH 03307, (603) 783-4931, Twitter: @NHMS
Phoenix International Raceway Cup Xfinity Truck	1.000	D-shaped Tri-Oval	Intermediate	7602 South Avondale Boulevard, Avondale, AZ 85323, www.phoenixraceway.com, (623) 463-5400, Twitter: @PhoenixRaceway

Pocono Raceway Cup Truck	2.500	Tri-oval	Superspeed-way	1234 Long Pond Rd, Long Pond, PA 18334, www.poconoraceway.com, (570) 646-2300, Twitter: @poconoraceway
Richmond International Raceway Cup Xfinity	0.750	D-shaped Oval	Short	600 E Laburnum Ave, Richmond, VA 23222, www.rir.com, (866) 455-7223, Twitter: @RIRInsider
Road America Xfinity	4.048	Road Course	Road Course	N7390 WI-67, Plymouth, WI 53073, (920) 892-4576, www.roadamerica.com, Twitter: @roadamerica
Sonoma Raceway Cup	1.990	Road Course	Road Course	29355 Arnold Dr, Sonoma, CA 95476, www.sonomaraceway.com, (707) 938-8448, Twitter: @RaceSonoma

Talladega Superspeedway Cup Xfinity Truck	2.660	Tri-oval	Superspeedway	3366 Speedway Blvd, Lincoln, AL 35096, www.talladegasuperspeedway.com, (877) 462-3342, Twitter: @TalladegaSuperS
Texas Motor Speedway Cup Xfinity Truck	1.500	Quad-oval	Intermediate	3545 Lone Star Cir, Fort Worth, TX 76177, www.texasmotorspeedway.com, (817) 215-8500, Twitter: @TXMotorSpeedway
Watkins Glen International Cup Xfinity	2.450	Road Course	Road Course	2790 County Route 16, Watkins Glen, NY 14891, www.theglen.com, (607) 535-2486, Twitter: @WGI

Major NASCAR Teams/Facilities/Shop Hours/Web/Social media info

Online map of NASCAR race shops: www.visitcabarrus.com/
cabarrus-county-map/?tid=93

BK Racing
11881 Vance Davis Drive
Charlotte, NC 28269
(704) 978-1022
Twitter: @BKRacing_2383
bkracingteam.com
Weekdays 9–noon and 1–5

Chip Ganassi Racing with Felix Sabates
8500 Westmoreland Drive
Concord, NC 28027
(704) 662-9642
Twitter: @CGRTeams
www.chipganassiracing.com
Weekdays 8–4:30

Circle Sport-Leavine Family Racing
6007 Victory Lane SW
Concord, NC 28027
(704) 455-7414
Twitter: @CSLFR95
lfr95.com
Call for hours

Dale Earnhardt Incorporated (Shop)
1675 Dale Earnhardt Highway 3
Mooresville, NC 28115

(704) 662-8000
Twitter: @DEIteam
www.daleearnhardtinc.com
Hours are limited (usually 11–2), call or check website.
Dale Earnhardt, Inc. is no longer a raceshop. It's now a museum/show-
room with a constantly changing Dale Earnhardt exhibit and gift shop
on 14 acres of land with 240,000 sq. feet of building space.

Front Row Motorsports
2670 Peachtree Road
Statesville, NC 28625
(704) 873-6445
Twitter: @Team_FRM
www.teamfrm.com
Weekdays 9–5

Furniture Row Racing
4000 Forest St
Denver, CO 80216
(303) 322-2008
Twitter: @FR78Racing
furniturerowracing.furniturerow.com
Weekdays 9–4

Germain Racing
218 Raceway Drive
Mooresville, NC 28117
(704) 799-4300
Twitter: @GEICORacing
germainracing.com
Weekdays 8–4
Tours available, limited availability.
Email: contact@germainracing13.com

Hendrick Motorsports
4400 Papa Joe Hendrick Blvd.
Charlotte, NC 28262
(877) 467-4890
Twitter: @TeamHendrick
www.hendrickmotorsports.com
Weekdays 8–4:30
Hendrick Motorsports is actually a campus that includes shops as well as the Hendrick Motorsports Museum & Team Store, all free.

Joe Gibbs Racing
13415 Reese Blvd. West
Huntersville, NC 28078
(704) 944-5000
Twitter: @JoeGibbsRacing
www.joegibbsracing.com
Viewing weekdays 8–5. Gift Shop 8–4:30

Joe Gibbs Racing. *Photo by Greg Engle*

JR Motorsports
349 Cayuga Drive
Mooresville, NC 28117
(704) 799-4800
Twitter: @JRMotorsports
www.jrmracing.com
Weekdays 8:30–5, Sat. 10–3, Sun. Closed
To request a shop tour email shoptours@jrmracing.com

JTG-Daugherty Racing
7201 Caldwell Road
Harrisburg, NC 28075
(704) 456-1221
Twitter: @NASCAR47
jtgdaughertyracing.com
Weekdays 8–5

NASCAR Hall of Fame
400 E. Martin Luther King, Jr. Blvd.
Charlotte, NC 28202
(704) 654-4400
Twitter: @NASCARHall
www.nascarhall.com
Seven days a week 10–6

Team Penske
200 Penske Way
Mooresville, NC 28115
(704) 664-2300
Twitter: @Team_Penske
www.teampenske.com
Weekdays 10–5

Visitors to the Team Penske shop will see one of the biggest shops in motorsports. There is a 440-foot fan walk above the garage floor and a 4,986-sq-ft gift shop.

Richard Childress Racing
425 Industrial Drive
Lexington, NC 27295
(336) 731-3334
Twitter: @RCRracing
www.rcrracing.com
Weekdays 10–5

Penske Racing. *Photo courtesy of Nigel Kinrade Photography for Penske Racing*

Richard Childress Racing Museum
236 Industrial Drive
Lexington, NC 27295
(800) 476-3389
Weekdays 10–4, Sat 10–2
www.rcrracing.com
Richard Childress Racing is about an hour north of Charlotte.

Richard Petty Motorsports
112 Byers Creek Road
Mooresville, NC 28117
(704) 743-5420
Twitter: @RPMotorsports
www.richardpettymotorsports.com
Weekdays, Saturdays 9–5

Roush Fenway Racing
4202 Roush Place
Concord, N.C. 28027
(704) 720-4600
Twitter: @roushfenway
www.roushfenway.com
Weekdays 8–5; Gift Shop: Mon-Fri 8:30–5
Race shops with museum, self-guided tour

Stewart-Haas Racing
6001 Haas Way
Kannapolis, N.C. 28081
(704) 652-4227
Twitter: @StewartHaasRcng
www.stewarthaasracing.com
Weekdays 8–4:30

Tommy Baldwin Racing
296 Cayuga Road
Mooresville, NC 28117
(704) 696-0036
Twitter: @TBR_Racing
tommybaldwinracing.com
Weekdays 9–4

Wood Brothers Racing
7201 Caldwell Road
Harrisburg, NC 28075
(704) 456-1421
Twitter: @woodbrothers21
www.woodbrothersracing.com
Weekdays 8:30–noon and 1–5

Driver Twitter Handles
(Current as of November 2016)

AJ Allmendinger: @AJDinger
Aric Almirola: @aric_almirola
Austin Dillon: @austindillon3
Ben Kennedy: @BenKennedy33
Benny Gordon: @BennyGordon24
Blake Koch: @BlakeKochRacing
Bobby Labonte: @Bobby_Labonte
Brad Coleman: @BradCColeman
Brad Keselowski: @keselowski
Brad Sweet: @BradSweet49
Brendan Gaughan: @Brendan62
Brennan Newberry: @brennannewberry
Brian Ickler: @BrianIckler
Brian Scott: @bscottracing
Brian Vickers: @BrianLVickers
Caleb Roark: @CalebRoark
Cameron Hayley: @CameronNHayley
Casey Mears: @CJMearsGang
Chad Boat: @ChadBoat
Chad McCumbee: @chad_mccumbee
Chase Elliott: @chaseelliott

Chase Miller: @ChaseMillerRCN

Clay Greenfield: @claygreenfield

Clint Bowyer: @ClintBowyer

Cole Whitt: @ColeWhitt

Coleman Pressley: @ColemanPressley

Colin Braun: @colinbraun

Corey LaJoie: @CoreyLaJoie

Dakoda Armstrong: @DakodaArmstrong

Dale Earnhardt Jr.: @DaleJr

Danica Patrick: @DanicaPatrick

Daniel Suárez: @Daniel_SuarezG

Darrell Wallace Jr: @BubbaWallace

David Gilliland: @DavidGilliland

David Ragan: @DavidRagan

David Starr: @starr_racing

Denny Hamlin: @dennyhamlin

Dylan Kwasniewski: @dylankracing

Elliott Sadler: @Elliott_Sadler

Eric McClure: @ericmcclure

Erik Darnell: @ErikDarnell_7

Grant Enfinger: @GrantEnfinger

Greg Biffle: @gbiffle

James Buescher: @JamesBuescher

Jamie McMurray: @jamiemcmurray

Jason Bowles: @Bowlesjason

Jason Leffler: @JasonLeffler

Jason White: @JasonWhite23

Jeb Burton: @JebBurtonRacing

Jeff Burton: @JeffBurton

Jeff Gordon: @JeffGordonWeb

Jeffrey Earnhardt @SlickEarnhardt

Jennifer Jo Cobb: @JenJoCobb

Jeremy Clements: @JClements51

Jimmie Johnson: @JimmieJohnson

JJ Yeley: @jjyeley1

Joe Nemechek: @FrontRowJoe87

Joey Coulter: @joeycoulter

Joey Gase Racing: @JoeyGaseRacing

Joey Logano: @joeylogano

Johnny Sauter: @JohnnySauter

Josh Wise: @Josh_Wise

Justin Allgaier: @J_Allgaier

Justin Lofton: @jlracing

Kasey Kahne: @kaseykahne

Kelly Bires: @KellyBires

Kevin Conway: @TheKevinConway

Kevin Harvick: @KevinHarvick

Kurt Busch: @KurtBusch

Kyle Busch: @KyleBusch

Kyle Fowler: @kylefowlerrace

Kyle Larson: @KyleLarsonRacin

Landon Cassill: @landoncassill

Mario Gosselin: @MarioGosselin

Martin Truex Jr.: @MartinTruex_Jr

Matt Crafton: @Matt_Crafton

Matt Kenseth: @mattkenseth

Matthew DiBenedetto: @mattdracing

Max Gresham: @MaxGresham

Michael Annett: @MichaelAnnett

Michael McDowell: @Mc_Driver

Michael Waltrip: @MW55

Michelle Tucker: @racerMichelleT

Mike Harmon: @hrmn8ter

Mike Skinner: @MSTheGunslinger

Paulie Harraka: @paulieharraka

Peyton Sellers: @PeytonSellers

Regan Smith: @ReganSmith
Ricky Stenhouse Jr.: @StenhouseJr
Ron Hornaday Jr.: @RonHornaday
Ross Chastain: @RossChastain
Ryan Blaney: @Blaney
Ryan Newman: @RyanJNewman
Ryan Sieg: @RyanSieg
Ryan Truex: @Ryan_Truex
Sam Hornish Jr.: @SamHornish
Steve Wallace: @stevewallace66
Tanner Berryhil: @tannerberryhill
Tayler Malsam: @TaylerMalsam
Timmy Hill: @TimmyHillRacer
Timothy Peters: @TimothyPeters17
TJ Bell: @TJBell_racing
Todd Bodine: @Team_Onion
Tony Stewart: @TonyStewart
Travis Kvapil: @TravisKvapil
Trevor Bayne: @Tbayne6
Ty Dillon: @tydillon
Ward Burton: @WardBurtonWBWF

Official NASCAR Twitter Accounts

NASCAR: @NASCAR
NASCAR XFINITY Series: @NASCAR_XFINITY
NASCAR Camping World Truck Series: @NASCAR_Trucks
NASCAR Touring and Weekly Series: @NASCARHomeTrack
IMSA: @UnitedSportsCar
NASCAR Stats: @NASCARStats
NASCAR Nation: @NASCARNATION
NASCAR Diversity: @NASCARDiversity
NASCAR Foundation: @NASCAR_FDN, @NASCARUnites

NASCAR Fuel For Business: @NASCARFFB

NASCAR Green: @NASCARGreen

NASCAR Hall of Fame: @NASCARHall

NASCAR Licensing: @NASCARMerch

NASCAR Performance: @NASCARauto

NASCAR Superstore: @SHOP_NASCAR

NASCAR Sprint Cup Series Sponsor: @MissSprintCup

NASCAR XFINITY Series Sponsor: @XFINITYRacing

Official NASCAR Executive Accounts

Steve Phelps (Executive VP and Chief Marketing Officer): @StevePhelps

Steve O'Donnell (Executive VP and Chief Racing Development Officer): @odsteve

Brent Dewar (Chief Operating Officer): @brentdewar

Jill Gregory (Senior Vice President and Chief Marketing Officer): @jillgregory6

Marc Jenkins (VP, Digital Media): @marcajenkins

Norris Scott (VP, Partnership Marketing and Business Solutions): @NorrisScott1

David Higdon (VP, Integrated Marketing Communications): @HigNASCAR

Jim Cassidy (Managing Director, Racing Operations): @jfcassidy

Zane Stoddard (VP, Entertainment Marketing): @zanestoddard

GLOSSARY

NASCAR Speak

200-mph tape: A type of super strong duct tape used to hold a damaged race car together long enough to finish the race. Usually used for smaller repairs or to hold down a piece of metal.

Aerodynamics: While this has many meanings outside NASCAR, when used in relation to motorsports, aerodynamics refers to the flow of air around, over, or under a race car at speed. NASCAR race cars are extensively tested in wind tunnels to help massage the surfaces and get the least amount of aerodynamic drag as possible. Many times the term is shortened to simply "aero."

A-post: This is the part of the car that goes from the roof to the base of the front windshield on both sides.

At Speed: In motorsports, "at speed" refers to the speed at which race cars travel during a green, or unrestricted, race condition.

Banking: The incline of a racetrack, most times at the turns, from the apron to the outside wall. Most tracks have some banking along the frontstretch, as well. Banking helps a car keep its speed and downforce while at speed. Talladega Superspeedway has 33 degrees of banking in its massive turns, 7 degrees along the backstretch,

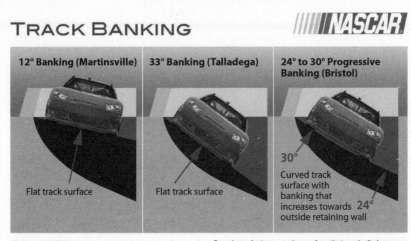

Banking. *Photo courtesy of NASCAR*

and 16.5 degrees along the frontstretch in the tri-oval. Indianapolis Motor Speedway has 7 degrees of banking in its 4 turns, making it one of the flattest. The degree of banking is measured at the height of the slope at the outside edge. Some tracks have progressive banking, meaning the banking is higher near the top than the bottom; this gives the track multiple grooves for racing, as a driver can go a bit faster near the top, while lower down the line is the shorter way around the track.

Bear Bond: Large sticky pieces of material used to make repairs to a damaged race car during a race.

B-post: The part of the car that goes from the roof to the base of the window behind a driver's head.

Bump and Run: A racing move where the car behind hits the car in front with enough force to cause the lead car to lose momentum. The car behind can then pass.

Camber: Refers to how much a tire is slanted in or out from straight up and usually refers to the front tires. This will always be described in degrees, either positive or negative. If the front wheels are set up properly, they will usually have the inside tire tilted out (positive) and the outside tire tilted in (negative). This will help a driver turn into the lefthand turns on ovals.

Commitment line: The line designating the point where a driver must "commit" to entering the pits. Once past this line, the driver must pit. If they don't, they face a commitment line violation from NASCAR.

Cut tire: This is a puncture or slice somewhere on the tire that causes it to lose air and fail. It can be caused by running over a piece of debris on the track or, as is often the case, contact with another racecar under race conditions.

Deck lid: Jargon for the part of the race car where the trunk lid is located.

Dirty air: An aerodynamic term, this refers to the unstable air generated by race cars at speed that can affect the race cars behind it. A race car normally performs better in "clean" air—that is, with nothing in front—than it does in dirty air, meaning behind another car.

Downforce: The amount of pressure from aerodynamic and centrifugal forces acting on a car. The more downforce, the better a race car will adhere to the track. There is a tradeoff, however, as too much downforce will add drag, which slows the car down.

Drafting: Another aerodynamic force, drafting is when two or more cars are racing single file, and almost touching. This will create a vacuum behind the lead car that will help pull the car behind it. The trailing car will help "push" the car in front, and together, both will go faster. Usually in play at the superspeedways, this is why you will see packs of cars at Talladega and Daytona running much faster than

NASCAR SPRINT CUP SERIES
AERODYNAMICS AND DRAFTING

Aerodynamics
Study of airflow in regard to a stock car.

Downforce
Downforce can be altered to improve the car's grip or traction by adjusting the spoiler as well as other aerodynamic changes to the car and its setup. As downforce is increased, the grip/traction is increased as well as tire wear. Increasing downforce comes at the expense of creating more drag, which will reduce fuel efficiency.

Draft
The aerodynamic effect that allows two or more cars traveling nose to tail to run faster than a single car. When one car follows another closely, the one in front punches through the air and provides a cleaner, less resistant path for the trailing cars.

Drag
The resistance a car experiences when passing through air at high speeds.

Drafting
The practice of two or more cars running nose to tail to create more speed for the group. The lead car displaces the air in front of it, creates a vacuum effect between its rear end and the nose of the second car and pulls the trailing cars along with it with less overall resistance. Two or more cars drafting will travel faster than a single car.

Source: NASCAR

Drafting. *Photo by NASCAR*

a lone car that has "lost the draft." Several years ago, NASCAR outlawed the practice of "bump-drafting," when the car behind bumps the car in front to gain a little bit more speed. Although not technically allowed, sometimes during a race two cars may try to gain an advantage with this practice, usually on the last lap of a race, when NASCAR is unlikely to penalize for it.

Drag: This is the aerodynamic force of resistance that a race car undergoes when racing. It slows a car down, as the air cannot easily move across the surface. The goal of teams is to build a car that is as "slick" as possible, meaning the air can move across the surface as smoothly as possible; any sort of drag prevents this.

Driving into the corner: A driver wants to get into a turn as smoothly and as quickly as possible. When he can't and has issues "driving into the corner," he could report that he is having a "tight" or "loose" condition.

Equalized: Tires used at speedways and superspeedways are actually two tires in one; there is an "inner" and "outer" liner. The inner liner has more air pressure than the outer liner. If the outer tire is cut during a race, the inner liner will help keep the tire up, allowing the driver to maintain control. When the inner liner (tire) loses air, the entire tire "equalizes" and creates a vibration.

Free Pass: A car, usually the Lucky Dog, who gets a lap back under caution.

Greenhouse: The area of a race car from the bottom of the windshield to the rear of the window in the back. Basically it's the area that in a streetcar would have the front and rear seats. In a race car, of course, there is only one seat; the greenhouse contains various other components such as the fire extinguisher, roll bars, and the driver's air conditioning system.

Green-White-Checkered Flag Finish: Also called an "overtime," this allows a race to go beyond its scheduled distance in an attempt to finish under green. The first lap is the restart (green), the next lap is the last lap (white), and when the field crosses the line, the checkered flag waves and the race ends. If a caution will make the race go past the scheduled distance, the field will be lined up and given the green flag. If the leader makes it to the "overtime" line, a line approximately halfway around the track, the next flag will end the race. If, however, a caution is waved prior to the leader reaching the overtime line, another attempt will be made at a green-white-checkered flag finish.

Groove: The quickest way around the racetrack. This can change depending on the driver, and there can be more than one groove, a high and low groove. Some drivers may prefer the high groove, others the low groove. Sometimes one groove may be faster, and drivers will often search for the faster groove. The groove can change during a race weekend as more drivers put rubber down (referring to taking laps during practice session and leaving parts of the tire), and as the weather changes. A groove will usually be very different during the day from what it is at night, and when the sun is shining on it, or when it is cloudy. A groove, upper or lower, can also appear as a race goes on. During a long race, you may hear a driver say that a groove "is coming in," meaning that it is developing. A groove will usually be evident as a dark lane around the track, especially in the turns.

Happy Hour: The final practice before the race, so called because crews and drivers need to be "happy" with their car when it's over because they will have no more time to make adjustments.

Hauler: A specially equipped 18-wheel tractor-trailer rig used to transport two race cars, engines, tools, and all the support equipment needed to race to the track. Cars are stored in the top section, while the bottom is used as a work area. Most team haulers are brightly decorated with sponsor colors and oftentimes a picture of the driver. NASCAR has its own hauler. Sometimes after an incident during a race, officials will call the offending parties to the hauler after the race; they are referring to

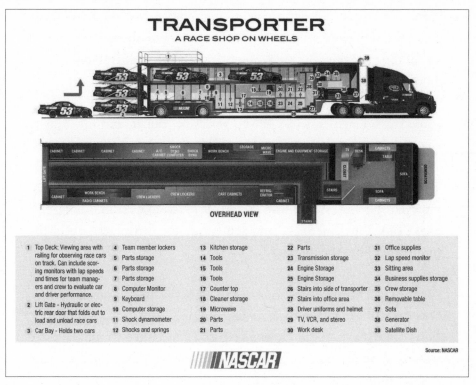

A NASCAR Hauler. *Photo courtesy of NASCAR*

the NASCAR hauler, which is like a student being called to the principal's office.

Lapped traffic: These are the race cars that are a full lap, or more, behind the leader of the race. A race leader will usually catch the back of the field under green flag conditions and must deal with them, since usually the leader is much faster.

Lead lap: The lap of record the race leader is currently on during a race.

Loading: Refers to the weight at a specified tire position on a race car due to aerodynamic downforce, the overall weight of the car, and lateral G-forces acting on it in a turn.

Loose: This is also called "oversteer." Loose refers to a condition when the rear tires of the car are having issues gripping in the corners and the rear of the car swings outward. Some drivers prefer a little bit of a loose condition at some tracks.

Loose stuff: Fragments like gravel, pieces of rubber, and other small bits of debris that collect near the outside wall or near the apron outside the groove. A driver who gets out of the groove is in danger of "getting into the loose stuff," which can cause the car to break traction, and even spin out. Either way, the driver must slow down. Loose stuff is also called the "marbles," and sometimes the driver gets out of the groove and "into the marbles."

Lucky Dog: During a race, the Lucky Dog is the first car one lap down. This car will get its lap back under caution, a free pass, as long as it didn't cause that caution. Drivers a lap down will often try and race one another in an attempt to be the first car one lap down, in the Lucky Dog position.

Lug nuts: These are the nuts that hold that tire on the wheel. NASCAR tires have five per tire, and they are put on and taken off using a high-pressure air wrench. NASCAR penalizes teams that don't use all five lug nuts during a race.

Pass Through: A penalty issued by NASCAR for such infractions as jumping the restart (passing before the leader reaches the start-finish line to take the green). A driver must come down pit road at pit road speed under racing conditions before returning to the track. Other common pit road penalties include entering the pit stall too soon (a driver can't drive through more than two stalls prior to his own), pitting outside the painted lines of the stall, crew members going over the wall too soon (prior to the stall before their cars), leaving with equipment still attached (a fuel can or a wrench still sticking out the back window), or an uncontrolled tire (a tire carrier or the crew member designated to catch it lets the tire get away and roll outside the pit box). NASCAR can also penalize a driver by sending him to the rear of the field or to the tail end of the longest line for such things as missing the prerace driver's meeting or driver introductions, or if the team makes adjustments to a car during an impound period.

Pole position: The number-one starting position. The pole is awarded to the driver who is the fastest qualifier, with the rest of the field lined up according to their individual qualifying effort. If a qualifying session can't be held (usually due to weather), NASCAR will "set the field" according to the current points.

Pulling up to pit: A term used by NASCAR to tell a driver they passed the pace car on the track as the driver was going to pit road. Under a caution, the pace car has control of the field. If the driver drops out of line to head to the pits, that driver cannot pass the pace car prior to getting to the line designating the entrance to the pits.

Restart: When the race is put back under a green flag (racing) condition. After a caution period, the race is restarted.

Restrictor plate: Currently used only at Daytona International Speedway and Talladega Superspeedway, a restrictor plate is a thin metal plate issued to teams by NASCAR that restricts the amount of airflow into the engine, sapping horsepower, and thus slowing the cars down. Once placed under carburetors, since NASCAR went to fuel injection in 2011, the restrictor plates are now placed under the throttle body. The races at Daytona and Talladega are sometimes referred to as "plate races" and the racing there as "plate racing."

RFID: This refers to the Radio Frequency Identification Chips that are installed on chassis at the NASCAR R&D Center in Concord,

North Carolina, after a race car has been certified by the sanctioning body. The RFID stores information showing how the chassis was set up when it was certified. At the track during inspection, the chip is scanned and the measurements compared. Officials can then determine if the car is within specifications. Goodyear also uses RFIDs to keep track of their tires.

Ride height: As it is in a passenger car, the ride height is the distance between the car's frame rails and the ground, or track.

Roll cage: This is the steel tubes seen inside a race car. The roll cage is designed to keep a driver safe in an accident or rollover. In NASCAR, there are very strict rules as to the placement, construction, and thickness of roll bars, and they are inspected on a regular basis.

Round: This is the way chassis adjustments are made using a race car's chassis springs. A wrench is inserted through a small hole in the rear window into a jack bolt attached to the springs. Putting "wedge" in, or taking wedge out, will tighten or lessen the pressure on the rear springs and loosen or tighten up the handling of a race car.

RPM: This acronym stands for "revolutions per minute" and is a measurement of the speed of the engine's crankshaft. Most NASCAR engines will run at or near 8000 rpms during a race.

Scuffs: Tires that have been on track and saved for later. Usually a lap or two is enough to "scuff" the tires in (each new tire has a sticker on it from the manufacturer on the tread surface; after a few laps, this sticker gets "scuffed" off, leaving only a few pieces). Most scuffs are used during qualifying, although sometimes in a race with a lot of tire changes, a team might use scuffs if their supply of new tires is getting low.

Setup: Refers to how a race car's suspension has been adjusted for the race. Adjustments can include shocks, springs, or air pressure in the tires.

Short pitting: A strategy where a team will pit before they need to, long before they would run out of fuel. Done under green flag conditions, the team will pit before a fuel window (the laps during a race when a race car needs fuel) opens. Once the rest of the field pits, the car that short-pitted will gain spots, even the lead. The rest of the race, the short-pitting car will be on a different cycle and could gain an advantage if no cautions fall.

Side Drafting: A move during a race where one car gets alongside another. This usually interrupts the airflow from the first car's nose and sends it to the rear of the first car. The first car then loses momentum, and the car that started the side draft will usually gain an advantage.

Slingshot: Legend has it that Junior Johnson discovered this move at Talladega and famously used it to win at the superspeedway many times. The car behind the leader in a draft shoots out and breaks the vacuum, giving the second car a burst of speed, and usually the lead. Many races at Talladega and Daytona have been won using the slingshot.

Splash and Go: A speedy pit stop where the race car gets only the fuel it needs to finish a race. The crew chief on the radio can oftentimes be heard counting down the seconds needed to get just enough fuel in the tank.

Splitter: A valance running along the front of the car that is near, or in some cases touching, the ground. The splitter helps with downforce in the front of the car. At a bumpy track, the splitter can sometimes become damaged if it is not set properly, and the race car will lose downforce.

Spoiler: A metal blade running along the rear deck lid of a race car. The spoiler is designed to "catch" some of the air, providing a bit of drag that pushes down the rear of the car for more downforce and traction.

Stickers: New tires. The name comes from the manufacturer's stickers that are found on each new tire's tread surface.

Stop and Go: This is a penalty used by NASCAR usually for speeding on pit road. The team must bring the car down pit road, and it must stop in its pit stall for one full second before heading back out to the track. No service can be done during a stop and go.

Tight: Also called "push" or "understeer." A driver reports their car as "tight" or pushing when the wheel is turned and the front end seems as though it's not turning. This is the opposite of a "loose" condition.

Toe: Different from camber, toe refers to the tires' top angle to one another. Toe-in means the top of the tires are pointing towards one another; toe-out is just the opposite.

Track bar: A long bar meant to keep the rear tires centered to the body of the car. The bar connects the car's frame on one side and the rear axle on the other. Also called the panhard bar, the driver can actually adjust the track bar from inside the car.

Trading paint: This describes aggressive racing with a lot of bumping and rubbing. Trading paint can sometimes end with both cars crashing.

Wave-around: These are the cars a lap down that can get a lap back when there is one to go prior to the green under caution, indicated by the pace car with its lights turned off. Any car planning to take a wave-around can't pit until after the green flag comes back out. Also, wave-around cars restart at the rear of the field, although in front of cars that have received a penalty.

Wedge: The cross-weight adjustment on a race car, wedge refers to adjusting the handling of the car by changing the pressure on the rear springs. Putting a round in will tighten the pressure, and taking a round out will lessen the pressure. This is often done during a pit stop when a crew member puts a special wrench through a small hole in the back window.

Window net: A woven mesh secured to the driver's side window, used to stop the driver's head and limbs from coming out during an accident. When a driver crashes, he signals to the safety workers that he is okay by unclipping and lowering the window net.

PAST NASCAR CHAMPIONS

2016 Jimmie Johnson	2000 Bobby Labonte
2015 Kyle Busch	1999 Dale Jarrett
2014 Kevin Harvick	1998 Jeff Gordon
2013 Jimmie Johnson	1997 Jeff Gordon
2012 Brad Keselowski	1996 Terry Labonte
2011 Tony Stewart	1995 Jeff Gordon
2010 Jimmie Johnson	1994 Dale Earnhardt
2009 Jimmie Johnson	1993 Dale Earnhardt
2008 Jimmie Johnson	1992 Alan Kulwicki
2007 Jimmie Johnson	1991 Dale Earnhardt
2006 Jimmie Johnson	1990 Dale Earnhardt
2005 Tony Stewart	1989 Rusty Wallace
2004 Kurt Busch	1988 Bill Elliott
2003 Matt Kenseth	1987 Dale Earnhardt
2002 Tony Stewart	1986 Dale Earnhardt
2001 Jeff Gordon	1985 Darrell Waltrip

1984 Terry Labonte

1983 Bobby Allison

1982 Darrell Waltrip

1981 Darrell Waltrip

1980 Dale Earnhardt

1979 Richard Petty

1978 Cale Yarborough

1977 Cale Yarborough

1976 Cale Yarborough

1975 Richard Petty

1974 Richard Petty

1973 Benny Parsons

1972 Richard Petty

1971 Richard Petty

1970 Bobby Isaac

1969 David Pearson

1968 David Pearson

1967 Richard Petty

1966 David Pearson

1965 Ned Jarrett

1964 Richard Petty

1963 Joe Weatherly

1962 Joe Weatherly

1961 Ned Jarrett

1960 Rex White

1959 Lee Petty

1958 Lee Petty

1957 Buck Baker

1956 Buck Baker

1955 Tim Flock

1954 Lee Petty

1953 Herb Thomas

1952 Tim Flock

1951 Herb Thomas

1950 Bill Rexford

1949 Red Byron

CITATIONS

Chapter 1 Citations

1. www. legendsofnascar.com/beginnings.htm
2. www. legendsofnascar.com/beginnings.htm
3. www. legendsofnascar.com/beginnings.htm
4. www.fireballroberts.com/Beginning.htm
5. NASCAR on television and radio, *World Heritage Encyclopedia*
6. About NASCAR: http://www.nascar.com/en_us/news-media/articles/about-nascar.html
7. Tracks get to work on upgrades: http://www.sportsbusinessdaily.com/Journal/Issues/2014/02/17/In-Depth/Tracks.aspx
8. Daytona Rising: http://www.daytonainternationalspeedway.com/Visitor-Center/DAYTONA-Rising.aspx

General history reference (timeline): www.nascarmedia.com/corpinfo/history.aspx

Bill France Sr., Betty Jane France: www.nascar.com/en_us/news-media/articles/2014/10/27/ups-game-changers-bill-france-sr-anne-bledsoe-france.html and hometracks.nascar.com/nascar_101/history_of_nascar

Ford 427 History: www.macsmotorcitygarage.com/2014/08/21/cammer-the-real-story-of-the-legendary-ford-427-sohc-v8/

Chapter 2 Citations

1. www.nascar.com/content/nascar/en_us/news-media/articles/about-nascar.html

2. www.nascar.com/content/nascar/en_us/news-media/articles/about-nascar.html

3. www.nascar.com/content/nascar/en_us/news-media/articles/about-nascar.html

4. www.courier-tribune.com/sports/pro/busch-caps-miraculous-career-first-points-title

5. www.gadsdentimes.com/sports/20140204/nascar-unveils-deterrence-penalty-system-for-2014

6. www.fireballroberts.com/Beginning.htm

INDEX